THE RELATIONS BETWEEN RELIGION AND SCIENCE

BY

FREDERICK TEMPLE

THE RELATIONS BETWEEN RELIGION AND SCIENCE

LECTURE I.

THE ORIGIN AND NATURE OF SCIENTIFIC BELIEF.

'O Lord, how manifold are Thy works: in wisdom hast Thou made them all; the earth is full of Thy riches.' —*Psalm* civ. 24.

Those who believe that the creation and government of the world are the work of a Being Whom it is their duty to love with all their hearts, Who loves them with a love beyond all other love, to Whom they look for guidance now and unending happiness hereafter, have a double motive for studying the forms and operations of Nature; because over and above whatever they may gain of the purest and highest pleasure in the study, and whatever men may gain of material comfort in a thousand forms from the results of the study, they cannot but have always present to their minds the thought, that all these things are revelations of His character, and to know them is in a very real measure to know Him. The believer in God, if he have the faculty and the opportunity, cannot find a more proper employment of time and labour and thought than the study of the ways in which God works and the things which God has made. Among religious men we ought to expect to find the most patient, the most truth-seeking, the most courageous of men of science.

We know that it is not always so; and that on the contrary Science and Religion seem very often to be the most determined foes to each other that can be found. The scientific man often asserts that he cannot find God in Science; and the religious man often asserts that he cannot find Science in God. Each often believes himself to be in possession, if not of the whole truth, at any rate of all the truth that it is most important to possess. Science seems to despise religion; and religion to fear and condemn Science. Religion, which certainly ought to put truth at the highest, is charged with refusing to acknowledge truth that has been proved. And Science, which certainly ought to insist on demonstrating every assertion which it makes, is charged with giving the rein to the imagination and treating the merest speculations as well-established facts.

To propose to reconcile these opposites would be a task which hardly any sane man would undertake. It would imply a claim to be able to rise at once above both, and see the truth which included all that both could teach. But it is a very useful undertaking, and not beyond the reach of thoughtful inquiry by an ordinary man, to examine the relations between the two, and thus to help not a few to find a way for themselves out of the perplexity. And this inquiry may well begin by asking what is the origin and nature of scientific belief on the one hand and of religious belief on the other. In this Lecture I propose to deal with the former.

It is not necessary to include in the Science of which I am to speak either Mathematics or Metaphysics. In as far as I need touch on what belongs to either, it will be only for the purpose of answering objections or of excluding what is irrelevant. And the consequent restriction of our consideration to the Science which concerns itself with Nature greatly simplifies the task that I have undertaken. For it will be at once admitted in the present day by all but a very few that the source of all scientific knowledge of this kind is to be found in the observations of the senses, including under that word both the bodily senses which tell us all we know of things external, and that internal sense by which we know all or nearly all that takes place within the mind itself. And so also will it be admitted that the Supreme Postulate, without which scientific knowledge is impossible, is the Uniformity of Nature.

Science lays claim to no revelations. No voice of authority declares what substances there are in the world, what are the properties of those substances, what are the effects and operations of those properties. No traditions handed down from past ages can do anything more than transmit to us observations made in those times, which, so far as we can trust them, we may add to the observations made in our own times. The materials in short which Science has to handle are obtained by experience.

But on the other hand Science can deal with these materials only on the condition that they are reducible to invariable laws. If any observation made by the senses is not capable of being brought under the laws which are found to govern all other observations, it is not yet brought under the dominion of Science. It is not yet explained, nor understood. As far as Science is concerned, it may be called as yet non-existent. It is for this very reason possible that the examination of it may be of the very greatest importance. To explain what has hitherto received no explanation constitutes the very essence of scientific progress. The observation may be imperfect, and may at once become explicable as soon as it is made complete; or, what is of far more value, it may be an instance of the operation of a new law not previously known, modifying and perhaps absorbing the law up to that time accepted. When it was first noticed in Galileo's time that water would not ascend in the suction pipe of a pump to a greater height than 32 feet, the old law that nature abhors a vacuum was modified, and the reasons why and the conditions under which Nature abhors a vacuum were discovered. The suction of fluids was brought under the general law of mechanical pressure. The doctrine that Nature abhorred a vacuum had been a fair generalization and expression of the facts of this kind that up to that time had been observed. A new fact was observed which would not fall under the rule. The examination of this fact led to the old rule being superseded; and Science advanced a great step at once. So in our own day was the planet Neptune discovered by the observation of certain facts which could not be squared with the facts previously observed unless the Law of Gravitation was to be corrected. The result in this case was not the discovery of a new Law but of a new Planet; and consequently a great confirmation of the old Law. But in each case and in every similar case the investigation of the newly observed fact proceeds on the assumption that Nature will be found uniform, and on no other assumption can Science proceed at all.

Now it is this assumption which must be first examined. What is its source? What is its justification? What, if any, are its limits?

It is not an assumption that belongs to Science only. It is in some form or other at the bottom of all our daily life. We eat our food on the assumption that it will nourish us to-day as it nourished us yesterday. We deal with our neighbours in the belief that we may safely trust those now whom we have trusted and safely trusted heretofore. We never take a journey without assuming that wood and iron will hold a carriage together, that wheels will roll upon axles, that steam will expand and drive the piston of an engine, that porters and stokers and engine-drivers will do their accustomed duties. Our crops are sown in the belief that the earth will work its usual chemistry, that heat and light and rain will come in their turn and have their usual effects, and the harvest will be ready for our gathering in the autumn. Look on while a man is tried for his life before a jury. Every tittle of the evidence is valued both by the judge and jury according to its agreement or disagreement with what we believe to be the laws of Nature, and if a witness asserts that something happened which, as far as we know, never happened at any other time since the world began, we set his evidence aside as incredible. And the prisoner is condemned if the facts before us, interpreted on the assumption that the ordinary laws of Nature have held their course, appear to prove his guilt.

What right have we to make such an assumption as this?

The question was first clearly put by Hume, and was handled by him with singular lucidity; but his answer, though very near the truth, was not so expressed as to set the question at rest.

The main relation in which the uniformity of Nature is observed is that of cause and effect. Hume examines this and maintains that there is absolutely nothing contained in it but the notion of invariable sequence. Two phenomena are invariably found connected together; the prior is spoken of as the cause, the posterior as the effect. But there is absolutely nothing in the former to define its relation to the latter, except that when the former is observed the latter, as far as we know, invariably follows. A ball hits another ball of equal size, both being free to move. There is nothing by which prior to experience we can determine what will happen next. It is just as conceivable that the moving ball should come back or should come to rest, as that the ball hitherto at rest should begin to move. A magnet fastened to a piece of wood is floating on water. Another magnet held in the hand is brought very near one of its poles or ends. If two north poles are thus brought together the floating magnet is repelled; if a north and a south pole are brought together the floating magnet is attracted. The motion of the floating magnet is in each case called the effect; the approach of the magnet held in the hand is called the cause. And this cause is, as far as we know, invariably followed by this effect. But to say that one is cause and the other effect is merely to say that one is always followed by the other; and no other meaning, according to Hume, can be attached to the words cause and effect.

Having established this interpretation of these words, Hume goes on to ask: What can be the ground in reason for the principle universally adopted, that the law of cause and effect rules phenomena, and that a cause which has been followed by an effect once will be followed by the same effect always? And he concludes that no rational ground can be found at all, that it is the mere result of custom without anything rational behind it. We are accustomed to see it so, and what we have been so perpetually accustomed to see we believe that we shall continue to see. But why what has always been hitherto should always be hereafter, no reason whatever can be given. The logical conclusion obviously is to discredit all human faculties and to land us in universal scepticism.

It was at this point that Kant took up the question, avowedly in consequence of Hume's reasoning. He considered that Hume had been misled by turning his attention to Physics, and that his own good sense would have saved him from his conclusion had he thought rather of Mathematics. Kant's solution of the problem, based mainly on the reality of Mathematics, and especially of Geometry, is the direct opposite of Hume's.

It will be most easy to give a clear account of Kant's solution by using a very familiar illustration. There is a well-known common toy called a Kaleidoscope, in which bits of coloured glass placed at one end are seen through a small round hole at the other. The bits of glass are not arranged in any order whatever, and by shaking the instrument may be rearranged again and again indefinitely and still without any order whatever. But however they may be arranged in themselves they always form, as seen from the other end, a symmetrical pattern. The pattern indeed varies with every shake of the instrument and consequent re-arrangement of the bits of glass, but it is invariably symmetrical. Now the symmetry in this case is not in the bits of glass; the colours are there no doubt, but the symmetrical arrangement of them is not. The symmetry is entirely due to the instrument. And if a competent enquirer looks into the instrument and examines its construction, he will be able to lay down with absolute certainty the laws of that symmetry which every pattern as seen through the instrument must obey.

Just such an instrument, according to Kant, is the human mind. Space and Time and the Perceptive Faculties are the parts of the instrument. Everything that reaches the senses must submit to the laws of Space and Time, that is, to the Laws of Mathematics, because Space and Time are forms of the mind itself, and, like the kaleidoscope, arrange all things on their way to the senses according to a pattern of their own. This pattern is as it were super-added to the manifestations that come from the things themselves; and if there be any manifestations of such a nature that they could not submit to this addition, or, in other words, could not submit to Mathematical Laws, these manifestations could not affect our senses at all. So too our Understanding has a pattern of its own which it imposes on all things that reach its power of perception. What cannot be accommodated to this pattern cannot be understood at all. Whatever things may be in themselves, their manifestations are not within the range of our intelligence, except by passing through the arranging process which our own mind executes upon them.

It is clear that this wonderfully ingenious speculation rests its claims for acceptance purely on the assertion that it and it alone explains the facts. It cannot be proved from any principle of reason. It assumes that there is a demonstrative science of Mathematics quite independent of experience, and that there are necessary principles of Physics equally independent of experience. And it accounts for the existence of these.

With Mathematics we are not now concerned, and I will pass them by with only one remark. The ground on which Kant's theory stands is not sufficient, for this simple reason. It accounts for one fact; it does not account for another fact. It accounts for the fact that we attach and cannot help attaching a conviction of necessity to all mathematical reasoning. We not only know that two straight lines cannot enclose a space, but we know that this is so and must be so in all places and at all times, and we know it without any proof whatever. This fact Kant accounts for. Space is according to him a part of our kaleidoscope; you can always look into it and see for yourself what are the laws of it. But there is another fact. This space of which we are speaking is unquestionably to our minds not a thing inside of us but outside of us. We are in it. We cannot get rid of a sense that it is independent of ourselves. We can imagine ourselves non-existing, minds and all. We cannot imagine space non-existing. If it be a part of our minds, how is it that we can picture to ourselves the non-existence of the mind which is the whole, but not the non-existence of space which, according to the hypothesis, is the part? For this fact, which we commonly call the objectivity of space, Kant's theory does not account. In fact Kant appears to have no escape from assigning this objectivity of space to delusion. But a theory which requires us to call an ineradicable conviction of consciousness a delusion cannot be said to explain all the facts. John Stuart Mill maintains that the other fact, namely, the conviction of the necessity of mathematical truth, is a delusion. And his account also must be pronounced for that reason to fail in accounting for all the facts.

But our present concern is not with Mathematics but with Physics. And here Kant fails altogether to convince; for, taking Time and the Perceptive Powers of the Understanding as parts of the human mind, he shows, what indeed is clearer and clearer every day, that the principles (so called) of Physics are indispensable Postulates, not indeed of observing with the senses, but of comprehending with the understanding, whatever happens. In order to give anything that can be called an explanation of any event we must show that it falls under the general rules which constitute the uniformity of Nature. We have no other meaning for the words understanding or explaining an event. Thinking, when analysed, is found to consist in bringing all that happens under universal laws, and no phenomenon can be said to be explained in thought except by being so related to all other phenomena. But it does not by any means follow that events cannot happen or cannot affect our senses without being susceptible of such explanation. To say that an event cannot be understood, and to say either that it cannot happen or that it cannot be observed by the senses, are two very different things. The fact is that Mathematics and Physics do not, as Kant assumes, present the same problem for solution, and do not therefore admit of one solution applicable to both. It is not the case that there is a science of abstract Physics corresponding to the science of Mathematics and sharing in the same character of necessity. In Mathematics we have truths which we cannot but accept, and accept as universal and necessary: in Physics we have no such truths, nor has Kant even endeavoured to prove that we have. The very question therefore that we are asked to solve in regard to Mathematics does not present itself in Physics. I am constrained to believe that two and two are four and not five; I am not constrained to believe that if one event is followed by another a great many times it will be so followed always. And the question is, why, without any constraint, I nevertheless so far believe it that I require special evidence in any given case to convince me to the contrary. And Kant's answer is irrelevant. He says that we cannot think the sequence of events unless they fall under the postulates of thinking, that is, the postulates of science; but this is no answer to the question. Why do we believe that, unless the contrary be proved, everything that is observed by the senses is capable of being reduced under these postulates of thinking? The sequence of things cannot otherwise be explained; but why should the sequence of all things that happen be capable of being explained? The question therefore still remains unanswered. What right have we to assume this Uniformity in Nature? or, in other words, what right have we to assume that all phenomena in Nature, observed by our senses, are capable of being brought within the domain of Science? And to answer this question we must approach it from a different side.

And there is the more reason for this because it is undeniable that both the definition and the universality of the relation of cause and effect, as they were accepted by Hume and his followers, are not accepted by men in general. In ordinary language something more is meant by cause and effect than invariable sequence, and the common assumption is not that all Nature obeys this rule with absolutely no variation, but that the rule is sufficiently general for all practical purposes.

If then we begin by asking what is the process of Science in dealing with all questions of causation, we find that this process when reduced to its simplest elements always consists in referring every event as an effect to some cause which we know or believe to have produced some other and similar event. Newton is struck by a falling apple. His first thought is, 'how hard the blow.' His second is wonder, 'how far the earth's attraction, which has caused this hard blow, extends.' His third, 'why not as far as the moon?' And he proceeds to assign the motion of the moon to the same cause as that which produced the motion of the apple. Taking this as a working hypothesis, he examines what would be the motions of all the planets if this were true. And the examination ends with establishing the high probability of the Law of Gravitation.

Now this being the invariable process of Science, it follows that our conception of cause must come originally from that cause which we have within ourselves and with which we cannot but begin, the action of the human will. It is from this action that is obtained that conception which underlies the ordinary conception of cause, namely, that of force or power.

This conception of force or power is derived from the consciousness of our own power to move our limbs, and perhaps too of passions, temptations, sentiments to move or oppose our wills. This power is most distinctly felt when it is resisted. The effort which is necessary when we choose to do what we have barely strength to do, impresses on us more clearly the sense of a force residing in ourselves capable of overcoming resistance. Having the power to move our limbs, and that too against some resistance, we explain, and in no other way can we explain, other motions by the supposition of a similar power. In so doing we are following strictly the scientific instinct and the scientific process. We are putting into the same class the motions that we observe in other things and the motions that we observe in ourselves; the latter are due to acts of our own wills, the former are assigned to similar acts of other wills. Hence in infancy, and in the infancy of mankind, the whole world is peopled with persons because everything that we observe to move is personified. A secret will moves the wind, the sun, the moon, the stars, and each is independent of the others.

Soon a distinction grows up between the things that seem to have a spontaneous motion and those that have not, and spontaneous motion is taken as the sign of life. And all inanimate things, of whatever kind, are held to be moved, if they move at all, by a force outside themselves. Their own force is limited to that of resisting, and does not include that of originating motion. But though they cannot originate motion they are observed to be capable of transmitting it. And the notion of force is expanded by the recognition that it can be communicated from one thing to another and yet to another, and that we may have to go back many steps before we arrive at the will from which it originated. We began with the notion of a power the action of which was or appeared to be self-originated: we come to the notion of a power the action of which is nothing more than the continuance of preceding action. And the special characteristic of the action of this force as thus conceived, which we may call the derivative force, is seen to be its regularity, just as the special characteristic of the self-originating action was its spontaneity.

As experience increases the regularity of the action of the derivative force is more and more observable, and then arises the notion of a law or rule regulating the action of every such force. And a perpetually increasing number of phenomena are brought under this head, and are shown to be, not the immediate results of self-originating action, but the more or less remote results of derivative action governed by laws. And even a large number of those phenomena, which specially belong to life and living creatures, in whom alone, if anywhere, the self-originating action is to be found, are observed to be subject to law and therefore to be the issue not of self-originating but of derivative action. And this observed regularity it is found possible to trace much more widely than it is possible to trace any clear evidence of what we understand by force. And so, at last, we frequently use the word force as it were by anticipation, not to express the cause of the phenomena, which indeed we do not yet know, but as a convenient abbreviation for a large number of facts classed under one head. And this it is which enables Hume to maintain that we mean no more by a cause than an event which is invariably followed by another event. We discover invariability much faster than we can discover causation; and having discovered invariability in any given case, we presume causation even when we cannot yet show it, and use language in accordance with that presumption. Thus, for instance, we speak of the force of gravitation, although we cannot yet prove that there is any such force, and all that we know is that material particles move as if such a force were acting on them.

As Science advances it is seen that the regularity of phenomena is far more important to us than their causes. And the attention of all students of Nature is fixed on that rather than on causation. And this regularity is seen to be more and more widely pervading all phenomena of every class, until the mind is forced to conceive the possibility that it may be absolutely universal, and that even will itself may come within its supreme dominion.

But to the very last the idea of causation retains the traces of its origin. For in the first place every step in this building up of science assumes a permanence underlying all phenomena. We cannot believe that the future will be like the past except because we believe that there is something permanent which was in the past and will be in the future. And this assumption of something permanent in things around us comes from the consciousness of something permanent within us. We know our own permanence. Whatever else we know or do not know about ourselves, we are sure of our own personal identity through successive periods of life. And as our explanation of things outside begins by classing them with things inside we still continue to ascribe permanence to whatever underlies phenomena even when we have long ceased to ascribe individual wills to any except beings like ourselves. And without this assumption of permanence our whole science would come to the ground.

And in the second place let it be remembered that we began with the will causing the motions of the limbs. Now there is, as far as we know, no other power in us to affect external nature than by setting something in motion. We can move our limbs, and by so doing move other things, and by so doing avail ourselves of the laws of Nature to produce remoter effects. But, except by originating motion, we cannot act at all. And, accordingly, throughout all science the attempt is made to reduce all phenomena to motions. Sounds, colours, heat, chemical action, electricity, we are perpetually endeavouring to reduce to vibrations or undulations, that is, to motion of some sort or other. The mind seems to find a satisfaction when a change of whatever kind is shown to be, or possibly to be, the result of movement. And so too all laws of Nature are then felt to be satisfactorily explained when they can be traced to some force exhibited in the movement of material particles. The law of Gravitation has an enormous evidence in support of it considered simply as a fact. And yet how many attempts have been made to represent it as the result of vortices or of particles streaming in all directions and pressing any two bodies together that lie in their path! The facts which establish it are enough. Why then these attempts? What is felt to be yet wanting? What is felt to be wanting is something to show that it is the result of some sort of general or universal motion, and that it thus falls under the same head as other motions, either those which originate in ourselves and are propagated from our bodies to external objects, or those which, springing from an unknown beginning, are for ever continuing as before.

This then is the answer to the question, Why do we believe in the uniformity of Nature? We believe in it because we find it so. Millions on millions of observations concur in exhibiting this uniformity. And the longer our observation of Nature goes on, the greater do we find the extent of it. Things that once seemed irregular are now known to be regular. Things that seemed inexplicable on this hypothesis are now explained. Every day seems to add not merely to the instances but to the wide-ranging classes of phenomena that come under the rule. We had reason long ago to hold that the quantity of matter was invariable. We now have reason to think that the quantity of force acting on matter is invariable. And to this is to be added the evidence of scientific prediction, the range of which is perpetually increasing, and which would be obviously impossible if Nature were not uniform. And yet again to this is to be added that this uniformity does not consist in a vast number of separate and independent laws, but that these laws already form a system with one another, and that that system is daily becoming more complete. We believe in the uniformity of Nature because, as far as we can observe it, that is the character of Nature.

And I use the word character on purpose, because it indicates better than any other word that I could find at once the nature and limitation of our belief.

For, if the origin of this belief be what I have described, it is perfectly clear that, however vast may be the evidence to prove this uniformity, the conclusion can never go beyond the limits of this evidence, and generality can never be confounded with universality. The certainty that Nature is uniform is not at all, and never can be, a certainty of the same kind as the certainty that four times five are twenty.

We can assert that the general character of Nature is uniformity, but we cannot go beyond this. Every separate law of nature is established by induction from the facts, and so too is the general uniformity. Every separate law of Nature is a working hypothesis. So too is the uniformity of Nature a working hypothesis, and it never can be more. It is true that there is far more evidence for the uniformity of Nature as a whole than for any one law of Nature; because a law of Nature is established by the uniformity of sequences in those phenomena to which it applies; whereas every uniformity of sequence, of whatever kind, is an evidence of the general uniformity. The evidence for the uniformity of nature is the accumulated evidence for all the separate uniformities. But, however much greater the quantity of evidence, the kind ever remains the same. There is no means by which we can demonstrate this uniformity. We can only make it probable. We can say that in almost every case all the evidence is one way; but whenever there is evidence to the contrary we cannot refuse to examine it.

If a miracle were worked science could not prove that it was a miracle, nor of course prove that it was not a miracle. To prove it to be a miracle would require not a vast range of knowledge, but absolutely universal knowledge, which it is entirely beyond our faculties to attain. To say that any event was a miracle would be to say that we knew that there was no higher law that could explain it, and this we could not say unless we knew all laws: to say that it was not a miracle would be *ex hypothesi* to assert what was false. In fact, to assert the occurrence of a miracle is simply to go back to the beginning of science, and to say: Here is an event which we cannot assign to that derivative action to which we have been led to assign the great body of events; we cannot explain it except by referring it to direct and spontaneous action, to a will like our own will. Science has shown that the vast majority of events are due to derivative action regulated by laws. Here is an event which cannot be so explained, any more than the action of our own free will can be so explained. Science may fairly claim to have shown that miracles, if they happen at all, are exceedingly rare. To demonstrate that they never happen at all is impossible, from the very nature of the evidence on which Science rests. But for the same reason Science can never in its character of Science admit that a miracle has happened. Science can only admit that, so far as the evidence goes, an event has happened which lies outside its province.

To believers the progress of Science is a perpetual instruction in the character which God has impressed on His works. That He has put Order in the very first place may be a surprise to us; but it can only be a surprise. In the great machinery of the Universe it constantly happens to us to find that that which is made indispensable, is nevertheless not the highest. The chosen people were not the highest in all moral or even in all spiritual characteristics; if we refuse the explanation given by Goethe that they were chosen for their toughness, yet we have no better to give. The eternal moral law is of all we know the highest and holiest. Yet the religious instinct seems to have been more indispensable for the development of humanity according to the Divine purpose than the observance of that moral law in all its fulness. It would never have occurred to us beforehand to permit in Divine legislation any concession to the hardness of men's hearts; yet we know that it was done. Science now tells us that Order takes a rank in God's work far above where we should have placed it. It is not the highest; it is far from the highest: but it appears to be in some strange way the most indispensable. God is teaching us that Order is far more universal, far more penetrating than we should have supposed. But, nevertheless, it is not itself God; nor the highest revelation of God. It is the stamp which, for reasons higher than itself, He appears to have put on His works. What is the limit to its application we do not know. There may be instances where this Order is apparently broken, but really maintained, because one physical law is absorbed in a higher; there may be instances where the physical law is superseded by a moral law. But we shall neither refuse to recognise that God has stamped this character on His works, nor let it on the other hand come between us and Him. For we know still that He is greater than all that He hath made, and He speaks to us by another voice besides the voice of Science.

LECTURE II.

THE ORIGIN AND NATURE OF RELIGIOUS BELIEF.

'So God created man in His own image, in the image of God created He him.' *Genesis* i. 27.

The order of phenomena is not the highest revelation of God, nor is the voice of Science the only nor the most commanding voice that speaks to us about Him. The belief in Him and in the character which we assign to Him does not spring from any observation of phenomena, but from the declaration made to us through the spiritual faculty.

There is within us a voice which tells of a supreme Law unchanged throughout all space and all time; which speaks with an authority entirely its own; which finds corroboration in the revelations of Science, but which never relies on those revelations as its primary or its ultimate sanction; which is no inference from observations by the senses external or internal, but a direct communication from the spiritual kingdom, the kingdom, as philosophers call it, of things in themselves; which commands belief as a duty, and by necessary consequence ever leaves it possible to disbelieve; and in listening to which we are rightly said to walk not by sight but by faith.

Now, before going on to say anything more about the message thus given to us from the spiritual world, it is necessary to consider an objection that meets us on the threshold of all such doctrines, namely, that it is simply impossible for us to know anything whatever of things in themselves. Our knowledge, it is urged, is necessarily relative to ourselves, whereas absolute as distinct from relative knowledge is for ever beyond our reach. We can speak of what things appear to us to be; we cannot speak of what they are. We know or may know whatever comes under the observation of our senses as phenomena; we cannot know what underlies these phenomena. And sometimes it has been maintained that we not only cannot know what it is that underlies the phenomena, but cannot even know whether anything at all underlies the phenomena, and that, for aught we can tell, the whole world and all that exists or happens in it may be nothing but a system of appearances with no substance whatever. This doctrine of the relativity of all knowledge is not only applied to things external but to our very selves. We know ourselves, it is maintained, only through an internal sense which can only tell us how we appear to ourselves, but cannot tell us in any the least degree what we really are.

Now this contention is an instance of a tendency against which we are required to be perpetually on our guard. The final aim of all science and of all philosophy is to find some unity or unities that shall co-ordinate the immense complexity of the world in which we live. Now there is one and only one legitimate way of attaining this aim, and that is by patient, persevering study of the facts. But the facts turn out to be so numerous, so multifarious, that not one life nor one generation but many lives and many generations will assuredly not co-ordinate them sufficiently to bring this aim within probable reach. Hence the incessant temptation, first, to supply by hypothesis what cannot yet be obtained by observation, and, secondly, to bend facts to suit this hypothesis; and, if the framing of such hypotheses be legitimate, the distortion of facts is clearly not legitimate. It seems too long to wait for future ages to complete the task. We must in some sort complete it now; and for that purpose if the facts as we observe them will not suit, we must substitute other facts that will. Accordingly every doctrine must be made complete, and to make this doctrine of the relativity of knowledge complete, we must get rid of all exceptions. But there is one exception that we cannot get rid of, and that is the conviction of our own identity through all changes through which we pass. Every man amongst us passes through incessant changes. His body changes; he may even lose parts of it altogether; he may lose all control over some of his limbs, or over them all. And there are internal as well as external changes in each man. His affections change, his practices, his passions, his resolutions, his purposes, his judgments; everything possibly by which he knows his own character. But through all these changes he is conscious of being still one and the same self. And he knows this; and knows it, not as an inference from any observation of sense external or internal, but directly and intuitively. All other knowledge may conceivably be relative, a knowledge of things as they appear, not of things in themselves. But this is not; it is a knowledge of a thing as it is in itself; for amidst all changes in the phenomena of each man's nature, this still remains absolutely unchanged. We do speak of sameness in application to phenomena; we say this is the same colour as that; this is the same musical note as that; this is the same sensation as that. But here we mean a different thing by the word same. We mean indistinguishability. We mean that we cannot distinguish between the two colours, the two notes, the two sensations. And this no doubt is a relative knowledge, not a knowledge of things in themselves. But we do not mean incapacity of being distinguished when we speak of our own personal identity. When a man thinks to-day of his life of yesterday, and regards himself as the same being through, all the time, he does not simply mean that he cannot distinguish between the being that existed yesterday according to his memory and the being that exists to-day according to his present consciousness: he means that the being is one and the same absolutely and in itself.

And this conviction of personal identity will presently be found to fall in with the revelation of the Moral Law, which is my subject in this Lecture. For it is by virtue of this personal identity that I become responsible for my actions. I am not merely the same thinking subject, I am the same moral agent all through my life. If I changed as fast as the phenomena of my being changed, my responsibility for any evil deed would cease the moment the deed was done. No punishment would be just, because it would not be just to punish one being for the faults of a totally different being. The Moral Law in its application to man requires as a basis the personal identity of each man with himself.

If corroboration were needed of the directness of the intuition by which we get this idea of our own personal identity, it would be found in the entire failure of all attempts to derive that idea from any other source. Comte, the founder of the Positive School, can do nothing with this idea but suggest that it is probably the result of some obscure synergy or co-operation of the faculties. John Stuart Mill passes it by altogether as lying outside the scope of his enquiries and of his doctrine. Mr. Herbert Spencer deals with it in a very weak chapter[1] of his remarkable volume of First Principles. He divides all the manifestations made to our consciousness, or, as we commonly say, all our sensations, into two great classes. He selects as the main but not universal characteristic of the one class, vividness; of the other class, faintness; a distinction first insisted on, though somewhat differently applied, by Hume. He adds various other characteristics of each class, some of them implying very questionable propositions. And we come finally to the following astonishing result. Sensations are divided into two classes; each has seven main characteristics which distinguish it from the other. One of these classes make up the subject, that which I mean when I use the words I myself; the other the object or that which is not I. But there is absolutely nothing to determine which is which, which class is the subject and which is the object, which is I myself, and which is not I myself. Vividness and faintness plainly have nothing in them by which we can assign the one to that which is I, the other to that which is not I. If we were to conjecture, we should be disposed to say that surely the most vivid sensations must be the nearest and therefore must be part of that which is I; but we find it is quite the other way. The faint sensations are characteristic of that which is I, and the vivid of that which is not I. And the same remark applies to each pair of characteristics in succession. The fact is that Mr. Spencer has omitted what is essential to complete his argument; he has not shown, nor endeavoured to show, nor even thought of showing, how out of his seven characteristics of the subject the conception of a subject has grown. It is quite plain that he not only makes his classes first and finds his characteristics afterwards, which we may admit to have been inevitable; but he fails altogether to show how that by which we know the classes apart has grown out of the characteristics that he has given us. The characteristics which he assigns to that which is I, all added together, do not in the slightest degree account for that sense of permanent existence in spite of changes which lies at the root of my distinction of myself from other things. The very word same, in the sense in which I use it when speaking of myself, cannot be defined except by reference to my own sameness with myself. It is a simple idea incapable of analysis, and is indeed, as was pointed out in my last Lecture, the root of the character of permanence which we assign to things external. To say that this conception has been evolved from the characteristics that Mr. Spencer has enumerated is like saying that a cat has been evolved without any intermediate stages from a fish, or a smell from a colour.

But, if we now go a step further, and ask in what form this personal identity presents itself in the world of phenomena, the answer is clear: our personality while bound up with all our other faculties, so that we can speak of our understanding, our affections, our powers of perception and sensation, as parts of ourselves, yet is centred in one faculty which we call the will. 'If there be aught spiritual in man,' says Coleridge, 'the will must be such. If there be a will, there must be a spirituality in man.' The will is the man. It is the will that makes us responsible beings. It is for the action of our will, or the consent of our will, that we come to be called in question. It is by the will that we assert ourselves amidst the existences around us; and as the will is the man in relation to phenomena, so on the other side the will is the one and only force among the forces of this world which takes cognizance of principles and is capable of acting in pursuit of an aim not to be found among phenomena at all. The will is not the whole spiritual faculty. Besides the power of willing we have the power of recognising spiritual truth. And this power or faculty we commonly call the conscience. But the conscience is not a force. It has no power of acting except through the will. It receives and transmits the voice from the spiritual world, and the will is responsible so far as the conscience enlightens it. It is the will whereby the man takes his place in the world of phenomena.

It is then to the man, thus capable of appreciating a law superior in its nature to all phenomena and bearing within himself the conviction of a personal identity underlying all the changes that may be encountered and endured, that is revealed from within the command to live for a moral purpose and believe in the ultimate supremacy of the moral over the physical. The voice within gives this command in two forms; it commands our duty and it commands our faith. The voice gives no proof, appeals to no evidence, but speaks as having a right to command, and requires our obedience by virtue of its own inherent superiority.

Its first command we call duty. The voice within awakes a peculiar sentiment which, except towards its command, is never felt in our souls, the sentiment of reverence. And it commands the pursuit of that, whatever it may be, to which this sentiment of reverence attaches. This is the positive test by which we are to know what is ever to be our highest aim. And along with this there is a negative test by which we are perpetually to correct the other, namely, the test of universality. The moral law in its own nature admits of no exceptions. If a principle of action be derived from this law it has nothing to do with time, or place, or circumstances; it must hold good in the distant future, in planets or stars utterly remote, as fully as it holds good now and here.

This duty we can subdivide under four heads, accordingly as we apply it to our dealings with ourselves, with other moral and spiritual beings, with other creatures that can feel pleasure and pain, with things that are incapable of either. If we are thinking of ourselves only, duty consists in the pursuit of holiness, that is, in the absolute subjection of what does not demand reverence to that which does. It is plain that what deserves reverence in us is that which approaches most nearly to the moral law in character. The appetites, the affections, the passions, have each their own separate objects. They may be useful in the highest degree, but they cannot in themselves deserve reverence, for their objects are not the moral law; they must therefore be absolutely subordinated to the will and the conscience which have for their objects the very law itself. Holiness consists in the subjection of the whole being, not in act alone, but in feeling and desire as well, to the authority of conscience.

If we are thinking of other moral agents, duty prescribes strict and unfailing justice; and justice in its highest and purest form is love, the unfailing recognition of the fullest claims that can be made on us by all who share our own divine superiority: to love God above all else, and to love all spiritual beings as we love ourselves, this is duty in relation to other spiritual beings.

If we are thinking of creatures which, whether moral agents or not, are capable of pain and pleasure, our duty takes the form of goodness or tenderness. We have no right to inflict pain or even refuse pleasure unless, if the circumstances were reversed, we should be bound in conscience to be ready in our turn to bear the same infliction or refusal. The precept, Do as you would be done by, is here supreme, and it is to this class of duties that that precept applies, and the limits of our right to inflict pain on other creatures, whether rational or irrational, will be determined by this rule.

And, lower still, our duty to things that are incapable of all feeling is summed up in that knowledge of them and that use of them which makes them the fittest instruments of a moral life.

The sentiment of reverence is our guide in determining our duty, and the test of universality perpetually comes in to correct the commands of this sentiment and to clear and so to refine the sentiment itself.

As is the case in a certain degree with every other kind of knowledge or belief, so in a very special degree the Moral Law finds its place even in minds that have very little of thought or of cultivation. The most untutored is not insensible to the claim made on our respect by acts of courage, self-sacrifice, generosity, truth; or to the call upon us for reprobation at the sight of acts of falsehood, of meanness, of cruelty, of profligacy. Even in the most untutored there is a sense that these sentiments of respect and reprobation are quite different in kind from the other sentiments which stir the soul. And this is even more clear in condemnation than in approval. However perverted the conscience (the seat of these sentiments) may be, yet the pain of remorse, which is self-reprobation for having broken the moral law, is always, as has been well said, 'quite unlike any other pain we know,' and is felt in some form and measure by every soul that lives. And as the sentiment thus holds a special place in the most untutored, so too does the sense of universality by which we instinctively and invariably correct or defend that sentiment if it be challenged. The moment we are perplexed in regard to what we ought to do or what judgment we ought to pass on something already done, we instinctively, almost involuntarily, endeavour to disentangle the act from all attendant circumstances and to see whether our sentiment of approval or disapproval would still hold good in quite other surroundings. We try to get, at the principle involved and to ascertain whether that principle possesses the universality which is the sure characteristic of the Moral Law.

It will be matter of consideration in a future Lecture how our knowledge of the Eternal Law of the holy, the just, the good, and the right, is thus purified in the individual and in the race. At present it will be enough to have indicated the general principle of what may be called the evolution of the knowledge of morals.

But I now go on from the Moral Law as a duty to the Moral Law as a faith. For the inner voice is not content with commanding a course of conduct and requiring obedience of that kind. This is its first utterance, and the man who hears and obeys unquestionably has within him the true seed of all religion. But though the first utterance it is not the last. For the same voice goes on to require us to believe that this Moral Law which claims obedience from us, equally claims obedience from all else that exists. It is absolutely supreme or it is nothing.

Its title to our obedience is its supremacy, and it has no other title. If it depended on promises of reward or threats of punishment addressed to us, it might be considered as a law for us, but could be no law for others. It would in that case, indeed, be a mere physical law. Things are so arranged for you, and as far as you know for you only, that terrible pain will come to you if you disobey, and wonderful pleasure if you obey. Such a law as that might proceed from a tyrant possessed of absolute power over US and the things that concern US, and might be either good or bad as should happen. But such a law would not be able to claim our reverence. Nay, rather, as is the case with all merely physical laws, it might be our duty to disobey it. In claiming our reverence as well as our obedience, in making its sanction consist in nothing but the fact of its own inherent majesty, the Moral Law calls on us to believe in its supremacy. It claims that it is the last and highest of all laws. The world before us is governed by uniformities as far as we can judge, but above and behind all these uniformities is the supreme uniformity, the eternal law of right and wrong, and all other laws, of whatever kind, must ultimately be harmonised by it alone. The Moral Law would be itself unjust if it bade us disregard all physical laws, and yet was itself subordinate to those physical laws. It has a right to require us to disregard everything but itself, if it be itself supreme; if not, its claim would be unjust. We see here in things around us no demonstrative proof that it is supreme, except what may be summed up in saying that there is a power that makes for righteousness. Enlightened by the Moral Law we can see strongly marked traces of its working in all things. The beauty, the order, the general tendency of all creation accords with the supremacy of the Moral Law over it all. But that is by no means all. We see, and we know that we see, but an infinitesimal fraction of the whole. And the result of this partial vision is that, while there is much in things around us which asserts, there is also much which seems to deny altogether any supremacy whatever in the Moral Law. The universe, as we see it, is not holy, nor just, nor good, nor right. The music of creation is full of discords as yet altogether unresolved. And if we look to phenomena alone, there is no solution of the great riddle. But in spite of all imperfections and contradictions, the voice within, without vouchsafing to give us any solution of the perplexity, or any sanction but its own authoritative command, imperatively requires us to believe that holiness is supreme over unholiness, and justice over injustice, and goodness over evil, and righteousness over unrighteousness. To obey this command and to believe this truth is Faith.

This is the Faith which is perpetually presenting to the believer's mind the vision of a world in which all the inequalities of this present world shall be redressed, in which truth, justice, and love shall visibly reign, in which temptations shall cease and sin shall cease also; in which the upward strivings of noble souls shall find their end, and holiness shall supersede penitence, and hearts shall be pure of all defilement. This is the Faith which holds to the sure conviction that all things shall one day come to judgment; and whether by sudden catastrophe or by sure development, the physical system shall surrender to the moral. This is the Faith which supplies perpetual strength to the hope of immortality; for though it cannot be said that the immortality of the individual soul is of necessity involved in a belief in the supremacy of the Moral Law, yet there is a sense, never without witness in the soul, that all would not be according to justice if a being to whom the Moral Law has been revealed from within is nevertheless in no degree to share in the final revelation of the superiority of that Moral Law over what is without. We cannot say that it is a necessary part of the supremacy of the Moral Law that every one of those who know it should partake of its immortal nature. We cannot even say that it is a necessary part of the ultimate redressing of all injustice and resolution of all the discords of life that the hope of it should prove true in the individual as it will certainly prove true in the universe. For we are unable to weigh individual merit or demerit, and cannot assert for certain that the balance of justice is not maintained even in this present life. But nevertheless the hope that it must and will be so is inextinguishable, and Faith in an Eternal Law of Morals is inextricably bound up with hope of immortality for the being that is endowed with a moral and responsible nature.

Faith in the absolute supremacy of the Moral Law is the first, but this again is not the last step upwards in Faith. We are called upon, and still by the same imperative voice within, to carry our Faith still further, and to believe something yet higher.

For the supremacy of the Moral Law must be a moral, not merely a physical supremacy. In claiming supremacy at all the Moral Law does not assert that somehow by a happy accident, as it were, all things turn out at last in accordance with what is in the highest sense moral. The supremacy of the moral over the physical involves in its very nature an intention to be supreme. It is not the supremacy of justice, if justice is done as the blind result of the working of machinery, even if that be the machinery of the universe. In our very conception of a moral supremacy is involved the conception of an intended supremacy. And the Moral Law in its government of the world reveals itself as possessing the distinctive mark of personality, that is, a purpose and a will. And thus, as we ponder it, this Eternal Law is shown to be the very Eternal Himself, the Almighty God. There is a sense in which we cannot ascribe personality to the Unknown Absolute Being; for our personality is of necessity compassed with limitations, and from these limitations we find it impossible to separate our conception of a person. And it will ever remain true that our highest conceptions of God must fall altogether short of His true nature. When we speak of Him as infinite, we are but denying that He is restrained by limits of time and space as we are. When we speak of Him as absolute, we are but denying that He is subject to conditions as we are. So when we speak of Him as a person, we cannot but acknowledge that His personality far transcends our conceptions. But it still remains the truth that these descriptions of Him are the nearest that we can get, and that for all the moral purposes of life we can argue from these as if they were the full truth. If to deny personality to Him is to assimilate Him to a blind and dead rule, we cannot but repudiate such denial altogether. If to deny personality to Him is to assert His incomprehensibility, we are ready at once to acknowledge our weakness and incapacity. But we dare not let go the truth that the holiness, the justice, the goodness, the righteousness, which the Eternal Moral Law imposes on us as a supreme command, are identical in essential substance in our minds and in His. Indeed, the more we keep before us the true character of that law, the more clearly do we see that the Moral Law is not His command but His nature. He does not make that law. He is that law. Almighty God and the Moral Law are different aspects of what is in itself one and the same. To hold fast to this is the fullest form of Faith. To live by duty is in itself rudimentary religion. To believe that the rule of duty is supreme over all the universe, is the first stage of Faith. To believe in Almighty God is the last and highest.

It will be seen at once by those who have followed me that I am in this Lecture only working out to its logical conclusion what was said long ago by Bishop Butler in England and by Kant in Germany. Butler calls the spiritual faculty whose commands to us I have been examining by the name of conscience: Kant calls it the practical reason. But both alike insist on the ultimate basis of morality being found in the voice within the soul and not in the phenomena observed by the senses. Science by searching cannot find out God. To reduce all the phenomena of the universe to order will not, even if it could ever be completely done, tell us the highest truth that we can attain to concerning spiritual things.

Science may examine all the phases through which religions have passed and treating human beliefs as it treats all other phenomena it can give us a history of religion or of religions. But there is something underlying them all which it cannot treat, and which perpetually evades all attempts to bring it under physical laws. For just as all attempts to explain away our conviction of our own personal identity have invariably failed and will for ever fail to satisfy human consciousness, so too the strictly spiritual element in all religion cannot be got out of phenomena at all. No analysis succeeds in obliterating the fundamental distinction between moral and physical law; or in enabling us to escape the ever increasing sense of the dignity of the former, or in shutting our ears to the still small voice which is totally unlike every other voice within or without. To bring the Moral Law under the dominion of Science and to treat the belief in it as nothing more than one of the phenomena of human nature, it is necessary to treat the sentiment of reverence which it excites, the remorse which follows on disobedience to its commands, the sense of its supremacy, as delusions. It is always possible so to treat these things; but only at the cost of standing lower in the scale of being.

But we have one step further to take. For as the spiritual faculty is the recipient directly or indirectly of that original revelation which God has made of Himself to His rational creatures, so too this appears to be the only faculty which can take cognizance of any fresh revelation that it might please Him to make. If He commands still further duties than those commanded by the supreme Moral Law, if He bids us believe what our reason cannot deduce from the primal belief in that Law and in Himself, it is to that faculty that the command is issued. If over and above the original religion as we may call it there is a revealed religion, it is the spiritual faculty that can alone accept it. Such a revelation may be confirmed by signs or proofs in the world of phenomena. He who is absolute over all nature may compel nature to bear witness to His teaching. The spiritual may burst through the natural on occasion, and that supremacy, which underlies all nature and which is necessarily visible to intelligences that are capable of seeing things as they are in themselves, may force itself into the world of phenomena and show itself in that manner to us. But this always is and must be secondary. The spiritual faculty alone can receive and judge of spiritual truth, and if that faculty be not reached a truly religious belief is not yet attained.

External evidences of revealed religion must have a high place but cannot have the highest. A revealed religion must depend for its permanent hold on our obedience and our duty on its fastening upon our spiritual nature, and if it cannot do that no evidences can maintain it in its place.

This account of the fundamental beliefs of Religion when compared with the fundamental postulates of Science shows that the two begin with the same part of our nature but proceed by opposite methods. Both begin with the human will as possessing a permanent identity and exerting a force of its own. But from this point they separate. Science rests on phenomena observed by the senses; Religion on the voice that speaks directly from the other world. Science postulates uniformity and is excluded wherever uniformity can be denied, but compels conviction within the range of its own postulate. Religion demands the submission of a free conscience, and uses no compulsion but that imposed by its own inherent dignity. Science gives warnings, and if you are capable of understanding scientific argument, you will be incapable of disbelieving the warnings. Certain things will poison you; certain neglects will ruin your health; disregard of scientific construction will bring your roof down on your head; to enter a burning building will risk your life; some of these things you may learn by ordinary experience, some of them by that combination of experience which is called Science. But if you are capable of the necessary reasoning you cannot doubt, however much you may wish to do so. And yet to defy these warnings and take the inevitable consequences of that defiance may be your highest glory. Religion also gives warnings; it assures you that the Eternal Moral Law is supreme; that, sooner or later, those who disobey will find their disobedience is exactly and justly punished; that no appearance to the contrary presented by experience can be trusted. But Religion will not compel you to believe any more than Science will compel you to obey. Disbelieve if you choose and Religion will do nothing but perpetually repeat its warnings and add that your disbelief has lowered you in the scale of being. So too Science gives promises; it promises, to the race rather than to the individual, life on easier conditions, and of greater length; fewer pains, fewer diseases; perpetually increasing comforts; perpetually increasing power over nature. And Science is sure to keep the promises. And yet we may refuse to accept the promises, and it is conceivable that the refusal may be far nobler than the acceptance. And Religion promises also. It promises stainless purity in the soul; and truth and justice and unfailing love; and tenderness to every creature that can feel; and a government of all that is under our dominion with a single eye to the service of God. And we may refuse to believe these promises or to care whether they are kept or not. But the refusal or pursuit of such aims as these determines our position in the judgment of the Supreme and in the court of our own conscience.

God has made man in His own image: that is, He has given man power to understand His works and to acknowledge Himself. And it is in acknowledging God that man finds himself divine. He is a partaker of the divine nature in proportion as he recognises the Supreme Law and makes it the law of his own will. And therefore has his will been made free as well as his mind rational: he has the power to choose as well as the power to know. And our choice lays hold on God Himself and makes us one with Him.

LECTURE III.

APPARENT CONFLICT BETWEEN SCIENCE AND RELIGION ON FREE-WILL.

'So God created man in His own image, in the image of God created He him.' *Genesis* i. 27.

Religion and Science both begin with the human will. The will is to Science the first example of power, the origin of the conception of cause; the bodily effort made by the will lies at the root of the conception of force. It is by comparing other forces with that force that Science begins its march. And the will is to religion the recipient of the Divine command. To the will the inner voice addresses itself, bidding it act and believe. It is because we have a will that we are responsible. In a world in which there were no creatures endowed with a will, there could be no right-doing or wrong-doing; no approval by conscience and no disapproval; no duty and no faith.

Here is the first possibility of collision between Science and Religion. Science postulates uniformity; Religion postulates liberty. Science cannot ever hope to reduce all phenomena to unity if a whole class of phenomena, all those that belong to the action of human will, are to be excluded from the postulate of invariable sequence. The action of the will is in this case for ever left outside. The evidence for the absolute uniformity of nature seems to be shaken, when it is found that there is so important a part of phenomena to which this law of uniformity cannot be applied. If a human will can thus interfere with the law of uniformity, there enters the possibility that behind some phenomena may lurk the interference of some other will. Religion, on the other hand, tells every man that he is responsible, and how can he be responsible if he is not free? If his action be determined by something which is not himself, how can the moral burden of it be put on him? To tell a man that he is to answer for it if he does something which he is tempted to do, is unmeaning, if he has no power to prevent himself from doing it.

But this is not all. For besides the sense of responsibility we have a direct consciousness of being free, a consciousness which no reasoning appears to extinguish. We sharply distinguish between that which goes on within us in regard to which we are free and that in regard to which we are not free. We cannot help being angry, but we can control our anger. We cannot help our wishes, but we can restrain our indulgence or our pursuit of them. We cannot directly determine our affections, but we can cherish or discourage them. There are extreme cases in which our wills seem powerless, but even here we are conscious of our power to struggle for self-assertion and self-control. There is very much in us which is not free; nay, there is much in us which impels us to action which is not free. But we never confound this with our wills, and when our wills are overpowered by passion or appetite, we call the act no longer a perfectly free act, and do not consider the responsibility for it to be quite the same.

This question of the freedom of the will was considered by Bishop Butler in the Analogy. He contented himself with proving that, make what theory we would concerning the necessity of human action, all men in practice acted on the theory of human freedom. We promise; we accept promises; we punish; we reward; we estimate character; we admire; we shun; we deal with ourselves; we deal with others; as if we and all others were free. And this was enough for his purpose. For he had to reconcile a Divine system of rewards and punishments with our sense of justice. And if he could show, as he did, that rewards and punishments were plainly not inconsistent with that sense of justice in our dealings with one another, it was impossible to call them inconsistent with that sense of justice in God's dealings with us.

But the purpose of these Lectures requires something more, and that for two reasons. For, in the first place, the doctrine of necessity was most often in Bishop Butler's days derived from a conception of a Divine foreknowledge arranging everything by supreme Will, not from the conception of a blind mechanical rule holding all in its unrelaxing grasp. And though to the cold reason it may make no difference how the will is bound, yet to the moral sentiment the two kinds of compulsion differ as life and death. To have no liberty because of being absolutely in the hands of Almighty God is quite another thing from having no liberty, as being under the dominion of a dead iron rule. It seems possible to accept the one and call it an unfathomable mystery; but to accept the other is to call life a delusion and the moral law a dream. And in the second place, the doctrine of necessity advanced as a theory and based on arguments not resting on facts, is a very different antagonist from the same doctrine advanced as a conclusion of science, and as deducible from a mass of co-ordinated observations. We may dismiss the mere theory after showing that it has not substance enough to hold its ground in ordinary life. We cannot so treat what claims to be a scientific inference.

The modern examination of the question begins with Hume, who maintains that the doctrine of liberty and that of necessity are both true and of course compatible with each other. But his arguments touch only the broad question whether they are true for practical purposes, not whether either is true in the strict sense and without exception or modification. To Kant's system, on the contrary, it was essential that both doctrines should be true in the strictest sense. Holding that invariable sequence was a law of Nature known independently of experience and applicable to all phenomena in the minutest detail, he could not allow that any act of the human will lay outside the range of this law. Such an act being a phenomenon must, in his view, be subject to the law which the constitution of our minds imposed on all phenomena apparent to us. And yet, on the other hand, holding that the eternal Moral Law made us responsible for all our acts, he could not but maintain that in the doing of those acts we must be free. His mode of reconciling the two opposites amounted to this, that our action throughout life considered as a whole is free, but that each separate act considered by itself is bound to the preceding acts by the law of invariable sequence. We may illustrate this by the familiar instance of a prism acting on a ray of light. The ray has or may have a colour of its own before it passes through the prism. The prism spreads it out and shows a series of colours. The order in which this series is arranged is determined by the character of the prism acting on the nature of the ray. The colours when combined give the colour of the ray; when separated by the prism each has its own distinct character, and the order of the colours is determined, and invariably determined, by the prism. So too in Kant's view the character of a man in itself may be free, but when it passes through the prism of time into the world of phenomena and is spread over many years it shows a number of separate actions, no one of which taken by itself exhibits the man, though all put together are the true representation of him to human perception. The man is free. His life represents his free choice. But his separate acts are what that free choice becomes when translated into a series of phenomena, and are bound each to the preceding by the law of invariable sequence. It is plain at once that this does not satisfy our consciousness. We are not conscious of freedom as regards our life as a whole; we are conscious of freedom as regards our separate actions. Our life as a whole embraces our past which is absolutely unchangeable, and our future which is not yet within our reach; we are conscious of no present power over either. Our separate acts are perceptibly subject to our own control; nay, it is by the use of our free-will in our separate acts that we are able to change the character of our life or to preserve it from change; and with this corresponds our responsibility. We hold ourselves responsible for each act as it is done; we hold ourselves responsible for the character of our lives only so far as we might have changed it by our acts. The solution leaves the difficulty where it was.

It is now customary with the advocates of the doctrine of necessity to express it by a different word, and call it the doctrine of determinism. The purpose of changing the word is to get rid of all associations with the idea of compulsion; just so in Science it is thought better to get rid of the words cause and effect, and substitute invariable sequence, in order to get rid of the notion of some compulsion recognisable by us in the cause to produce the effect. Determinism does not say to a man 'you will be forced to act in a particular way;' but 'you will assuredly do so.' There will be no compulsion; but the action is absolutely certain. Just as on a given day the moon will eclipse the sun, so in given circumstances you will do the precise thing which it is your character in such circumstances to do. And your sense of freedom is simply the sense that the action proceeds from yourself and not from any force put upon you from without.

But this too does not solve the problem. It is true that in regard to a very large proportion of our actions the sense of freedom seems to be no more than negative. We do what it is our custom, our inclination, our character to do. We are not conscious of any force being put upon us; but neither are we conscious of using any force ourselves. We float as it were down the stream, or hurry along with a determined aim, but having no desire nor purpose to the contrary, the question of freedom or necessity never seems to arise. It is even possible and common for us not to know ourselves as well as others know us, and to do many things which an observer would predict as sure to be our actions, but which we ourselves fancy to be by no means certain. Even in these cases we sometimes awake to the fact that what we are thus allowing in our lives is not consistent with the law of duty, and, do what we may, we cannot then escape the conviction that we are to blame, and that we had power to act otherwise if only we had chosen to exert the power. But it is when a conflict arises between duty and inclination that our inner certainty of our own freedom of will becomes clear and unconquerable. In the great conflicts of the soul between the call of duty and the power of temptation there are two forces at work upon us. We are never for a moment in doubt which is ourselves and which is not ourselves; which is the free agent and which is the blind force; which is responsible for the issue, and which is incapable of responsibility. There is in this case a real sense of compulsion from without, and a real sense of resistance to that compulsion from within. It is impossible in this case to account for the sense of being a free agent, by saying that this merely means that we are conscious of no external force. We are conscious of an external force and we are conscious that this will of ours which struggles against it is not an external force, but our very selves, and this distinction between the will and the forces against which the will is striving is ineffaceable from our minds. That the will is often weak and on that account overpowered, and that after a hard struggle our actions are often determined, not by our wills but by our passions or our appetites, is unquestionable. Often has the believer to pray to God for strength to hold fast to right purpose, and often will he feel that without that strength he must inevitably fall. But he knows that whatever source may supply the strength, it is he that will have to use it, and he that will be responsible for using it or neglecting to do so.

The advocates of determinism urge that every action must have a motive, and that the man always acts on that motive which is the stronger. The first proposition may be granted at once. The freedom of the will is certainly not shown in acting without any motive at all. If there be any human action which appears to be without any motive, it is not in such action that we find human freedom. Such action, if possible at all, must inevitably be mechanical. A man who is acting from mere caprice is even more completely at the mercy of passing inclination than one who is acting from passion or from overpowering temptation. The freedom of the will is not shown in acting without motive, but in choosing between motives. But when it is further said that a man always acts from the stronger motive, the question immediately follows, what determines which is the stronger motive? It cannot be anything in the motives themselves, or all men would act alike in the same circumstances; and it is clear that they do not. It must be therefore something in the man. And if it be something in the man, it must be either his will acting at the moment, which in that case is free, or his character. But if it be his character, then follows the further question, what determines his character? If we are to maintain the uniformity of nature, we must answer by assigning the determination to the sum total of surrounding and preceding circumstances. Nothing will satisfy that law of uniformity but this; that, given such and such parents, such and such circumstances of birth and life, there must be such a character and no other. At what point is there room in this case for any responsibility? I did not on this supposition make my character; it was made for me; any one else born in my stead, and living in my stead, would of necessity have acted exactly as I have done; would have felt the same, and aimed at the same, and won the same moral victories, and suffered the same moral defeats. How can I be held responsible for what is the pure result of the circumstances in which I was born? But if, on the other hand, it be said that our character is not the mere fruit of our antecedents and surroundings, the law of uniformity is clearly broken. A new element has come into the world, namely, my character, which has not come out of the antecedents and surroundings according to any fixed law. The antecedents and surroundings might have been quite the same for any one else, and yet I should have my character and he his, and our lives would have altogether differed.

It is clear that determinism does not get us out of the difficulty. Here, too, as in regard to the necessary truths of mathematics, and in regard to the relativity of all our knowledge, the theory has purchased completeness by the cheap expedient of calling one of the facts to be accounted for a delusion. Such a solution cannot be accepted. In spite of all attempts to explain it away, the fact that we think ourselves free and hold ourselves responsible remains, and remains unaffected.

But let us examine how far the difference between the scientific view and the religious view of human action extends.

Observation certainly shows that a very large proportion of human action, much even of that which appears at first sight to be more especially independent of all law, is really as much regulated by laws of nature as the movements of the planets. I have already pointed out how often an observer can predict a man's actions better than the man himself, and how often the will is certainly passive and consents instead of acting. In these cases there is no reason whatever to deny that nature and not the will is producing the conduct. And not only so, but that which seems most irregular, the kind of action that we call caprice, there is very often just as little reason to call free, as to assign free-will as the cause of the uncertainties of the weather. But it is not in observing individuals so much as in observing masses of men that we get convincing proof that men possess a common nature, and that their conduct is largely regulated by the laws of that nature. That amongst a given large number of men living on the whole in the same conditions from year to year, there should be every year a given number of suicides, of murderers, of thieves and criminals of various kinds, cannot be accounted for in any other way than by the hypothesis that like circumstances will produce like conduct. So, too, in this way only can we account for such a fact as the steadiness in the proportion of men who enter any given profession, of men who quit their country for another, of men who remain unmarried all their lives, of men who enter a university, of men who make any particular choice (such as these) which can be tested by figures. Now, this argument is unanswerable as far as it goes; but it succeeds, like all the other arguments for the uniformity of nature, in establishing the generality, and not at all the universality of that uniformity. Indeed, it falls far short of proving as much uniformity in human action as is proved in the action of inanimate things. The induction which proves the uniformity of the laws of mechanics, of chemistry, of physics, is so far greater than the induction which proves the uniformity of human conduct, that it is hardly possible to put the two side by side. When we turn from abstract arguments to facts, the doctrine of necessity is unquestionably unproven.

And this agrees with the result of a careful examination of the facts of human consciousness from the opposite point of view. We cannot but acknowledge that when we look very closely we find a very large proportion of our own actions to be by no means the result of an interference by the will. A large proportion is due to custom; a large proportion to inclination, of which the will takes no special notice, and is not called on by the conscience to notice; a large proportion to inclinations which we know that we ought to resist, but we do not resist; a much smaller proportion, but still some, to passions and appetites against which we have striven in vain; only a very small proportion to deliberate choice. There is, in fact, no irresistible reason for claiming freedom for human action except when that action turns on the question of right or wrong. There is no reason to call action free that flows from inclination or custom, or passion, or a desire to avoid pain, or a desire to obtain pleasure. The will claims to be free in all these cases, but it is free in the sense that it might be exerted; and so, since it is not exerted, the action is not free. But when, at the call of duty, in whatever form, the will directly interferes, then and then only are we conscious not only that the will is free, but that it has asserted its freedom, and that the action has been free also.

The relation of the will to the conduct falls under four distinct heads: for sometimes the will simply concurs with the inclination; sometimes it neither concurs nor opposes; sometimes it opposes but is overpowered; sometimes it opposes and prevails. In the first case, inclination of some kind or other prompts the man to action. The inclination, whether set up by an external object of desire or by an internal impulse of restlessness or blind craving or the like, comes clearly from the nature, and is not free choice. There is no reason to believe that it is not in most cases, possibly in all cases, under the dominion of fixed law. It may be as completely the product of what has preceded it as the eclipse of the sun. And if the will concurs in the inclination, it is needless to discuss the question whether the will acts or not. The conduct is the same whether the will adds force to the inclination or is simply passive. The freedom of the will may in this case be considered as negative. So, too, may the freedom of the will be considered negative in the second case, which is that of the will neither concurring with inclination nor opposing it. In this case there may be a distinct consciousness of freedom in the form of a sense of responsibility for what inclination is permitted to do. A man in this case knows that he is free, perhaps knows that he ought to interfere and control the conduct. But as he does not interfere, the freedom of the will is not asserted in act. And it is possible that, as far as all external phenomena are concerned, there may be no breach in uniformity of sequence. This, however, can hardly be in the third case, which is when the will and the inclination are opposed, and the will is overpowered. Although the inclination prevails, yet the struggle itself is an event of the most important kind, and is sure to leave traces on the character, and to be followed by consequences. In this case we are distinctly conscious of a power to add force to that one of the contending opposites which is most identified with our very selves, and we know whether we have added that force or not. And not only may we add this force directly from within; we may and we often do go outside of ourselves to seek for aids to add still more force indirectly, and we do for this purpose what we should not do otherwise. We dwell in thought on the higher aims which are the proper object of will; we read what sets forth those higher aims in their full beauty; we seek the words, the company, the sympathy of men who will, we are sure, encourage us in this the higher path. And, on the other hand, we turn away from the temptation which gives strength to the evil inclination, and if we cannot escape from its presence we endeavour to drive the thought of it from our minds. All this action is not for the sake of anything thus done, but for the sake of its indirect effect on the struggle in which we are engaged. Whenever there is a struggle, we are not only conscious that the will is free, but that it is asserting its freedom. In these struggles there is not a mere contest between two inclinations. We are distinctly conscious that one of the combatants is our very selves in a sense in which the other is not. But, nevertheless, when all has been said, it still remains in this case that the will is beaten and inclination prevails, and the conduct in the main is determined by the inclination, which is under the dominion of the law of uniformity, and not by the will, which claims to be free. The fourth case in which the will prevails may, of course, make a momentous breach in the uniformity of sequence of the conduct. But in far the largest number of cases the struggle is very slight, and the difference between the will and the inclination is not, taken alone, of grave importance in the life. And in those instances in which the struggle is severe and the resulting change is great, it is very often the case that the way has been prepared, as it were in secret, by the quiet accumulation of hidden forces of the strictly natural order ready to burst forth when the fit opportunity came. In the great conversions which have sometimes seemed by their suddenness and completeness to defy all possibility of reduction to natural law, there are often nevertheless tokens of deep dissatisfaction with the previous life having swelled up slowly within the soul for some time, even for some long time beforehand. The inclination to go on in evil courses has been broken down at last, not merely by the action of the will, but by the working of the machinery of the soul.

To this it must be added that the action of the will is such that it very often happens that, having been exerted once, it need not be exerted again for the same purpose. A custom is broken down, an exceedingly strong temptation has been overpowered, and its strength so destroyed that its return is without effect. Or sometimes the act of the will takes the form of deliberately so arranging the circumstances of life that a dreaded temptation cannot return, or if it return cannot prevail; the right eye has been plucked out, the right hand cut off, and the sin cannot be committed even if desired. While therefore the will is always free, the actual interference of the will with the life is not so frequent as to interfere with the broad general rule that the course of human conduct is practically uniform. In fact the will, though always free, only asserts its freedom by obeying duty in spite of inclination, by disregarding the uniformity of nature in order to maintain the higher uniformity of the Moral Law. The freedom of the human will is but the assertion in particular of that universal supremacy of the moral over the physical in the last resort, which is an essential part of the very essence of the Moral Law. The freedom of the will is the Moral Law breaking into the world of phenomena, and thus behind the free-will of man stands the power of God.

When the real claim of the will for freedom has been clearly seized by the mind, it becomes apparent that there is no real collision between what Science asserts and what Religion requires us to believe. Science asserts that there is evidence to show that an exceedingly large proportion of human action is governed by fixed law. Religion requires us to believe that the will is responsible for all this action, not because it does, but because it might interfere. Science is not able, and from the nature of the case never will be able to prove that the range of this fixed law is universal, and that the will never does interfere to vary the actions from what without the will they would have been. Science will never be able to prove this, because it could not be proved except by a universal induction, and a universal induction is impossible. At present there is no approximation to such proof. Religion, on the other hand, does not call on us to believe that the will often interferes, but on the contrary is perpetually telling us that it does not interfere as often as it ought. Revealed religion, indeed, has always based its most earnest exhortations on the reluctance of man to set his will to the difficult task of contending with the forces of his nature, and on the weakness of the will in the presence of those forces.

And when we pursue this thought further we see that for such creatures as we are the subjection of a large part of our own nature to fixed laws is as necessary for our dominion over ourselves as the fixity of external nature is necessary for our dominion over the world around us. The fixity of a large part of our nature—nay, of all but the whole of it—is a moral and spiritual necessity. For it requires but a superficial self-examination to discern the indications of what the profoundest research still leaves a mystery—that we are not perfect creatures of our own kind—that our nature does not spontaneously conform to the Supreme Moral Law—that our highest and best consists not in complete obedience to which we cannot attain, but in a perpetual upward struggle. Now such a struggle demands for its indispensable condition something fixed in our nature by which each step upwards shall be made good as it is taken, and afford a firm footing for the next ascent. If there were nothing in us fixed and firm, if the warfare with evil impulses, wayward affections, overmastering appetites had to be carried on through life without the possibility of making any victory complete, the formation of a perpetually higher and nobler character would be impossible; our main hope in this life, our best offering to God would be taken away from us; we could never give our bodies to be a living sacrifice, holy, acceptable to God; we could give our separate acts but not ourselves, for we should be utterly unable to form ourselves into fitness for such a purpose. The task given to the will is not only to govern the actions but to discipline the nature; but discipline is impossible where there is no fixity in the thing to be disciplined.

And this becomes still more important when we search more deeply and perceive that not the nature only but the will itself is in some strange way infected with evil. We can hardly imagine even a perfectly pure will capable of continuing to the end a conflict in which no progress ever was or could be made. The tremendous strain of fighting with an enemy that might be defeated again and again for ever without ever suffering any change or relaxing the violence of any attack or giving the slightest hope of any relief, would seem too much forthe most unearthly, the most noble, the most godlike of human wills. But wills such as ours, penetrated with weakness, perhaps with treachery to their own best aspirations, how utterly impossible that they could persevere through such a hopeless conflict.

It is the sustaining hope of the Christian that he shall be changed from glory to glory into the image or likeness of His Lord, and that when all is over for this life he shall be indeed like Him and see Him as He is. But that hope is never presented as one to be realized by some sudden stroke fashioning the soul anew and moulding it at once into heavenly lineaments. It is by steady and sure degrees that the Christian believes that he shall be thus blessed. And this progress rests on the fixed rules by which his nature is governed, and which admit of the character being gradually changed by the life. The Christian knows that God has so made us that a temptation once overcome is permanently weakened, and often overcome is at last altogether expelled; that appetites restrained are in the end subdued and cost but little effort to keep down; that bad thoughts perpetually put aside at last return no more; that a clearer perception of duty and a more resolute obedience to its call makes duty itself more attractive, fills us with enthusiasm for its fulfilment, draws us as it were upwards, and ennobles the whole man. The Christian knows that the thought of the Supreme Being, the contemplation of His excellency, the recognition of Him as the source of spiritual life has a strange power to transform, and evermore to transform the whole man. In this knowledge the Christian lives his life and fights his battle. And what is this but a knowledge that he has a nature subject to fixed laws, which he can indeed interfere with, but without which his self-discipline would be of little value, and assuredly could not long continue.

And if the progress of Science and the examination of human nature should eventually restrict more closely than we might have supposed the length to which the interference of the will can go; if it should appear that the changes which we can make at any one moment in ourselves are within a very narrow range, this, too, will be knowledge that can be used in our self-discipline and quite as much perhaps in our mutual moral aid. It is conceivable that the branch of science which treats of human nature may in the end profoundly modify our modes of education, and our hopes of what can be effected by it. But if so the knowledge will only add to the store of means put within our reach for the elevation of our race. And we may be sure that nothing of this sort will really affect the revelation that God has written in our souls that we are free and responsible beings, and cannot get quit of our responsibility.

LECTURE IV.

APPARENT CONFLICT BETWEEN RELIGION AND THE DOCTRINE OF EVOLUTION.

'For the invisible things of Him from the creation of the world are clearly seen, being understood by the things that are made, even His eternal power and Godhead.' *Romans* i. 20.

The regularity of nature is the first postulate of Science; but it requires the very slightest observation to show us that, along with this regularity, there exists a vast irregularity which Science can only deal with by exclusion from its province. The world as we see it is full of changes; and these changes when patiently and perseveringly examined are found to be subject to invariable or almost invariable laws. But the things themselves which thus change are as multifarious as the changes which they undergo. They vary infinitely in quantity, in qualities, in arrangement throughout space, possibly in arrangement throughout time. Take a single substance such, say, as gold. How much gold there is in the whole universe, and where it is situated, we not only have no knowledge, but can hardly be said to be on the way to have knowledge. Why its qualities are what they are, and why it alone possesses all these qualities; how long it has existed, and how long it will continue to exist, these questions we are unable to answer. The existence of the many forms of matter, the properties of each form, the distribution of each: all this Science must in the last resort assume.

But I say in the last resort. For it is possible, and Science soon makes it evident that it is true, that some forms of matter grow out of other forms. There are endless combinations. And the growth of new out of old forms is of necessity a sequence, and falls under the law of invariability of sequences, and becomes the subject-matter of Science. As in each separate case Science asserts each event of to-day to have followed by a law of invariable sequence on the events of yesterday; the earth has reached the precise point in its orbit now which was determined by the law of gravitation as applied to its motion at the point which it reached a moment ago; the weather of the present hour has come by meteorological laws out of the weather of the last hour; the crops and the flocks now found on the surface of the habitable earth are the necessary outcome of preceding harvests and preceding flocks and of all that has been done to maintain and increase them; so, too, if we look at the universe as a whole, the present condition of that whole is, if the scientific postulate of invariable sequence be admitted, and in as far as it is admitted, the necessary outcome of its former condition; and all the various forms of matter, whether living or inanimate, must for the same reason and with the same limitation be the necessary outcome of preceding forms of matter. This is the foundation of the doctrine of Evolution.

Now stated in this abstract form this doctrine will be, and indeed if Science be admitted at all must be, accepted by everybody. Even the Roman Church, which holds that God is perpetually interfering with the course of nature, either in the interests of religious truth or out of loving kindness to His creatures, yet will acknowledge that the number of such interferences almost disappears in comparison of the countless millions of instances in which there is no reason to believe in any interference at all. And if we look at the universe as a whole, the general proposition as stated above is quite unaffected by the infinitesimal exception which is to be made by a believer in frequent miracles. But when this proposition is applied in detail it at once introduces the possibility of an entirely new history of the material universe. For this universe as we see it is almost entirely made up of composite and not of simple substances. We have been able to analyse all the substances that we know into a comparatively small number of simple elements—some usually solid, some liquid, some gaseous. But these simple elements are rarely found uncombined with others; most of those which we meet with in a pure state have been taken out of combination and reduced to simplicity by human agency. The various metals that we ordinarily use are mostly found in a state of ore, and we do not generally obtain them pure except by smelting. The air we breathe, though not a compound, is a mixture. The water which is essential to our life is a compound. And, if we pass from inorganic to organic substances, all vegetables and animals are compound, sustained by various articles of food which go to make up their frames. Now, how have these compounds been formed? It is quite possible that some of them, or all of them to some extent, may have been formed from the first. If Science could go back to the beginning of all things, which it obviously cannot, it might find the composition already accomplished, and be compelled to start with it as a given fact—a fact as incapable of scientific explanation as the existence of matter at all. But, on the other hand, composition and decomposition is a matter of every-day experience. Our very food could not nourish us except by passing through these processes in our bodies; and by the same processes we prepare much of our food before consuming it. May not Science go back to the time when these processes had not yet begun? May not the starting-point of the history of the universe be a condition in which the simple elements were still uncombined? If Science could go back to the beginning of all things, might we not find all the elements of material things ready indeed for the action of the inherent forces which would presently unite them in an infinite variety of combinations, but as yet still separate from each other? Scattered through enormous regions of space, but drawn together by the force of gravitation; their original heat, whatever it may have been, increased by their mutual collision; made to act chemically on one another by such increase or by subsequent decrease of temperature; perpetually approaching nearer to the forms into which, by the incessant action of the same forces, the present universe has grown; these elements, and the working of the several laws of their own proper nature, may be enough to account scientifically for all the phenomena that we observe. We do not even then get back to regularity. Why these elements, and no others; why in these precise quantities; why so distributed in space; why endowed with these properties: still are questions which Science cannot answer, and there seems no reason to expect that any scientific answer will ever be possible. Nay, I know not whether it may not be asserted that the impossibility of answering one at least among these questions is capable of demonstration. For the whole system of things, as far as we know it, depends on the perpetual rotation of the heavenly bodies; and without original irregularity in the distribution of matter no motion of rotation could ever have spontaneously arisen. And if this irregularity be thus original, Science can give no account of it. Science, therefore, will have to begin with assuming certain facts for which it can never hope to account. But it *may* begin by assuming that, speaking roughly, the universe was always very much what we see it now, and that composition and decomposition have always nearly balanced each other, and that there have been from the beginning the same sun and moon and planets and stars in the sky, the same animals on the earth and in the seas, the same vegetation, the same minerals; and that though there have been incessant changes, and possibly all these changes in one general direction, yet these changes have never amounted to what would furnish a scientific explanation of the forms which matter has assumed. Or, on the other hand, Science *may* assert the possibility of going back to a far earlier condition of our material system; may assert that all the forms of matter have grown up under the action of laws and forces still at work; may take as the initial state of our universe one or many enormous clouds of gaseous matter, and endeavour to trace with more or less exactness how these gradually formed themselves into what we see. Science has lately leaned to the latter alternative. To a believer the alternative may be stated thus: We all distinguish between the original creation of the material world and the history of it ever since. And we have, nay all men have, been accustomed to assign to the original creation a great deal that Science is now disposed to assign to the history. But the distinction between the original creation and the subsequent history would still remain, and for ever remain, although the portion assigned to the one may be less, and that assigned to the other larger, than was formerly supposed. However far back Science may be able to push its beginning, there still must lie behind that beginning the original act of creation—creation not of matter only, but of the various kinds of matter, and of the laws governing all and each of those kinds, and of the distribution of this matter in space.

This application of the abstract doctrine of Evolution gives it an enormous and startling expansion: so enormous and so startling that the doctrine itself seems absolutely new. To say that the present grows by regular law out of the past is one thing; to say that it has grown out of a distant past in which as yet the present forms of life upon the earth, the present vegetation, the seas and islands and continents, the very planet itself, the sun and moon, were not yet made—and all this also by regular law—that is quite another thing. And the bearings of this new application of Science deserve study.

Now it seems quite plain that this doctrine of Evolution is in no sense whatever antagonistic to the teachings of Religion, though it may be, and that we shall have to consider afterwards, to the teachings of revelation. Why then should religious men independently of its relation to revelation shrink from it, as very many unquestionably do? The reason is that, whilst this doctrine leaves the truth of the existence and supremacy of God exactly where it was, it cuts away, or appears to cut away, some of the main arguments for that truth.

Now, in regard to the arguments whereby we have been accustomed to prove or to corroborate the existence of a Supreme Being, it is plain that, to take these arguments away or to make it impossible to use them, is not to disprove or take away the truth itself. We find every day instances of men resting their faith in a truth on some grounds which we know to be untenable, and we see what a terrible trial it sometimes is when they find out that this is so, and know not as yet on what other ground they are to take their stand. And some men succumb in the trial and lose their faith together with the argument which has hitherto supported it. But the truth still stands in spite of the failure of some to keep their belief in it, and in spite of the impossibility of supporting it by the old arguments.

And when men have become accustomed to rest their belief on new grounds the loss of the old arguments is never found to be a very serious matter. Belief in revelation has been shaken again and again by this very increase of knowledge. It was unquestionably a dreadful blow to many in the days of Galileo to find that the language of the Bible in regard to the movement of the earth and sun was not scientifically correct. It was a dreadful blow to many in the days of the Reformation to find that they had been misled by what they believed to be an infallible Church.

Such shocks to faith try the mettle of men's moral and spiritual conviction, and they often refuse altogether to hold what they can no longer establish by the arguments which have hitherto been to them the decisive, perhaps the sole decisive, proofs.

And yet in spite of these shocks belief in revelation is strong still in men's souls, and is clearly not yet going to quit the world.

But let us go on to consider how far it is true that the arguments which have hitherto been regarded as proving the existence of a Supreme Creator are really affected very gravely by this doctrine of Evolution.

The main argument, which at first appears to be thus set aside, is that which is founded on the marks of design, and which is worked out in his own way with marvellous skill by Paley in his Natural Theology. Paley's argument rests as is well known on the evidence of design in created things, and these evidences he chiefly finds in the frame-work of organised living creatures. He traces with much most interesting detail the many marvellous contrivances by which animals of various kinds are adapted to the circumstances in which they are to live, the mechanism which enables them to obtain their food, to preserve their species, to escape their enemies, to remove discomforts. All nature thus examined, and particularly all animated nature, seems full of means towards ends, and those ends invariably such as a beneficent Creator might well be supposed to have in view. And whilst there is undeniably one great objection to his whole argument, namely that the Creator is represented as an Artificer rather than a Creator, as overcoming difficulties which stood in His way rather than as an Almighty Being fashioning things according to His Will, yet the argument thus drawn from evidence of design remains exceedingly powerful, and it has always been considered a strong corroboration of the voice within which bids us believe in a God. Now it certainly seems at first as if this argument were altogether destroyed. If animals were not made as we see them, but evolved by natural law, still more if it appear that their wonderful adaptation to their surroundings is due to the influence of those surroundings, it might seem as if we could no longer speak of design as exhibited in their various organs; the organs we might say grow of themselves, some suitable, and some unsuitable to the life of the creatures to which they belonged, and the unsuitable have perished and the suitable have survived.

But Paley has supplied the clue to the answer. In his well-known illustration of the watch picked up on the heath by the passing traveller, he points out that the evidence of design is certainly not lessened if it be found that the watch was so constructed that, in course of time, it produced another watch like itself. He was thinking not of Evolution, but of the ordinary production of each generation of animals from the preceding. But his answer can be pushed a step further, and we may with equal justice remark that we should certainly not believe it a proof that the watch had come into existence without design if we found that it produced in course of time not merely another watch but a better. It would become more marvellous than ever if we found provision thus made not merely for the continuance of the species but for the perpetual improvement of the species. It is essential to animal life that the animal should be adapted to its circumstances; if besides provision for such adaptation in each generation we find provision for still better adaptation in future generations, how can it be said that the evidences of design are diminished? Or take any separate organ, such as the eye. It is impossible not to believe until it be disproved that the eye was intended to see with. We cannot say that light was made for the eye, because light subserves many other purposes besides that of enabling eyes to see. But that the eye was intended for light there is so strong a presumption that it cannot easily be rebutted. If indeed it could be shown that eyes fulfilled several other functions, or that species of animals which always lived in the dark still had fully-formed eyes, then we might say that the connexion between the eye of an animal and the light of heaven was accidental. But the contrary is notoriously the case; so much the case that some philosophers have maintained that the eye was formed by the need for seeing, a statement which I need take no trouble to refute, just as those who make it take no trouble to establish, I will not say its truth, but even its possibility. But the fact, if it be a fact, that the eye was not originally as well adapted to see with as it is now, and that the power of perceiving light and of things in the light grew by degrees, does not show, nor even tend to show, that the eye was not intended for seeing with.

The fact is that the doctrine of Evolution does not affect the substance of Paley's argument at all. The marks of design which he has pointed out remain marks of design still even if we accept the doctrine of Evolution to the full. What is touched by this doctrine is not the evidence of design but the mode in which the design was executed. Paley, no doubt, wrote on the supposition (and at that time it was hardly possible to admit any other supposition) that we must take animals to have come into existence very nearly such as we now know them: and his language, on the whole, was adapted to that supposition. But the language would rather need supplementing than changing to make it applicable to the supposition that animals were formed by Evolution. In the one case the execution follows the design by the effect of a direct act of creation; in the other case the design is worked out by a slow process. In the one case the Creator made the animals at once such as they now are; in the other case He impressed on certain particles of matter which, either at the beginning or at some point in the history of His creation He endowed with life, such inherent powers that in the ordinary course of time living creatures such as the present were developed. The creative power remains the same in either case; the design with which that creative power was exercised remains the same. He did not make the things, we may say; no, but He made them make themselves. And surely this rather adds than withdraws force from the great argument. It seems in itself something more majestic, something more befitting Him to Whom a thousand years are as one day and one day as a thousand years, thus to impress His Will once for all on His creation, and provide for all its countless variety by this one original impress, than by special acts of creation to be perpetually modifying what He had previously made. It has often been objected to Paley's argument, as I remarked before, that it represents the Almighty rather as an artificer than a creator, a workman dealing with somewhat intractable materials and showing marvellous skill in overcoming difficulties rather than a beneficent Being making all things in accordance with the purposes of His love. But this objection disappears when we put the argument into the shape which the doctrine of Evolution demands and look on the Almighty as creating the original elements of matter, determining their number and their properties, creating the law of gravitation whereby as seems probable the worlds have been formed, creating the various laws of chemical and physical action, by which inorganic substances have been combined, creating above all the law of life, the mysterious law which plainly contains such wonderful possibilities within itself, and thus providing for the ultimate development of all the many wonders of nature.

What conception of foresight and purpose can rise above that which imagines all history gathered as it were into one original creative act from which the infinite variety of the Universe has come and more is coming even yet?

And yet again, it is a common objection to Paley's and similar arguments that, in spite of all the tokens of intelligence and beneficence in the creation, there is so much of the contrary character. How much there is of apparently needless pain and waste! And John Stuart Mill has urged that either we must suppose the Creator wanting in omnipotence or wanting in kindness to have left His creation so imperfect. The answer usually given is that our knowledge is partial, and, could we see the whole, the objection would probably disappear. But what force and clearness is given to this answer by the doctrine of Evolution which tells us that we are looking at a work which is not yet finished, and that the imperfections are a necessary part of a large design the general outlines of which we may already trace, but the ultimate issue of which, with all its details, is still beyond our perception! The imperfections are like the imperfections of a half-completed picture not yet ready to be seen; they are like the bud which will presently be a beautiful flower, or the larva of a beautiful and gorgeous insect; they are like the imperfections in the moral character of a saint who nevertheless is changing from glory to glory.

To the many partial designs which Paley's Natural Theology points out, and which still remain what they were, the doctrine of Evolution adds the design of a perpetual progress. Things are so arranged that animals are perpetually better adapted to the life they have to live. The very phrase which we commonly use to sum up Darwin's teaching, the survival of the fittest, implies a perpetual diminution of pain and increase of enjoyment for all creatures that can feel. If they are fitter for their surroundings, most certainly they will find life easier to live. And, as if to mark still more plainly the beneficence of the whole work, the less developed creatures, as we have every reason to believe, are less sensible of pain and pleasure; so that enjoyment appears to grow with the capacity for enjoyment, and suffering diminishes as sensitivity to suffering increases. And there can be no doubt that this is in many ways the tendency of nature. Beasts of prey are diminishing; life is easier for man and easier for all animals that are under his care: many species of animals perish as man fills and subjugates the globe, but those that remain have far greater happiness in their lives. In fact, all the purposes which Paley traces in the formation of living creatures are not only fulfilled by what the Creator has done, but are better fulfilled from age to age. And though the progress may be exceedingly slow, the nature of the progress cannot be mistaken.

If the Natural Theology were now to be written, the stress of the argument would be put on a different place. Instead of insisting wholly or mainly on the wonderful adaptation of means to ends in the structure of living animals and plants, we should look rather to the original properties impressed on matter from the beginning, and on the beneficent consequences that have flowed from those properties. We should dwell on the peculiar properties that must be inherent in the molecules of the original elements to cause such results to follow from their action and reaction on one another. We should dwell on the part played in the Universe by the properties of oxygen, the great purifier, and one of the great heat-givers; of carbon, the chief light-giver and heat-giver; of water, the great solvent and the storehouse of heat; of the atmosphere and the vapours in it, the protector of the earth which it surrounds. We should trace the beneficent effects of pain and pleasure in their subservience to the purification of life. The marks of a purpose impressed from the first on all creation would be even more visible than ever before.

And we could not overlook the beauty of Nature and of all created things as part of that purpose coming in many cases out of that very survival of the fittest of which Darwin has spoken, and yet a distinct object in itself. For this beauty there is no need in the economy of nature whatever. The beauty of the starry heavens, which so impressed the mind of Kant that he put it by the side of the Moral Law as proving the existence of a Creator, is not wanted either for the evolution of the world or for the preservation of living creatures. Our enjoyment of it is a super-added gift certainly not necessary for the existence or the continuance of our species. The beauty of flowers, according to the teaching of the doctrine of Evolution, has generally grown out of the need which makes it good for plants to attract insects. The insects carry the pollen from flower to flower, and thus as it were mix the breed; and this produces the stronger plants which outlive the competition of the rest. The plants, therefore, which are most conspicuous gain an advantage by attracting insects most. That successive generations of flowers should thus show brighter and brighter colours is intelligible. But the beauty of flowers is far more than mere conspicuousness of colours even though that be the main ingredient. Why should the wonderful grace, and delicacy, and harmony of tint be added? Is all this mere chance? Is all this superfluity pervading the whole world and perpetually supplying to the highest of living creatures, and that too in a real proportion to his superiority, the most refined and elevating of pleasures, an accident without any purpose at all? If Evolution has produced the world such as we see and all its endless beauty, it has bestowed on our own dwelling-place in lavish abundance and in marvellous perfection that on which men spend their substance without stint, that which they value above all but downright necessities, that which they admire beyond all except the Law of Duty itself. We cannot think that this is not designed, nor that the Artist who produced it was blind to what was coming out of His work.

Once more, the doctrine of Evolution restores to the science of Nature the unity which we should expect in the creation of God. Paley's argument proved design, but included the possibility of many designers. Not one design, but many separate designs, all no doubt of the same character, but all worked out independently of one another, is the picture that he puts before us. But the doctrine of Evolution binds all existing things on earth into one. Every mineral, every plant, every animal has such properties that it benefits other things beside itself and derives benefit in turn. The insect developes the plant, and the plant the insect; the brute aids in the evolution of the man, and the man in that of the brute. All things are embraced in one great design beginning with the very creation. He who uses the doctrine of Evolution to prove that no intelligence planned the world, is undertaking the self-contradictory task of showing that a great machine has no purpose by tracing in detail the marvellous complexity of its parts, and the still more marvellous precision with which all work together to produce a common result.

To conclude, the doctrine of Evolution leaves the argument for an intelligent Creator and Governor of the world stronger than it was before. There is still as much as ever the proof of an intelligent purpose pervading all creation. The difference is that the execution of that purpose belongs more to the original act of creation, less to acts of government since. There is more divine foresight, there is less divine interposition; and whatever has been taken from the latter has been added to the former.

Some scientific students of Nature may fancy they can deduce in the working out of the theory results inconsistent with religious belief; and in a future Lecture these will have to be examined; and it is possible that the theory may be so presented as to be inconsistent with the teaching of Revelation. But whatever may be the relation of the doctrine of Evolution to Revelation, it cannot be said that this doctrine is antagonistic to Religion in its essence. The progress of Science in this direction will assuredly end in helping men to believe with more assurance than ever that the Lord by wisdom hath founded the earth, by understanding hath He established the heavens.

LECTURE V.

REVELATION THE MEANS OF DEVELOPING AND COMPLETING SPIRITUAL KNOWLEDGE.

'God, who at sundry times and in divers manners spake in time past to the Fathers by the Prophets, hath in these last days spoken to us by His Son.'*Hebrews* i. 1.

The doctrine of Evolution has been applied not only to the formation of all created things, but to the development of human knowledge; and this with perfect justice, though with some risk of misunderstanding. It is certain, and, indeed, it is obvious, that knowledge grows. The ordinary experience of mankind becomes larger and clearer in the course of time, and the systematised experience which we call Science makes the same progress in still greater measure and with more assurance.

Our Science has been built on the labours of scientific men in past ages. New generalisations imagined by one thinker, new crucial experiments devised by another, new instruments of observation invented by another,—these have been the steps by which Science has grown and established its authority and enlarged its dominion. When or by whom the first steps were made we have no record. No mathematician that ever lived showed greater natural power of intellect than he, whoever he was, who first saw that the singular contained the universal; but we know neither his name nor his age, nor his birthplace nor his race. But after those first steps had been taken, we know who have been the leaders in scientific advance. And we know what they have done, and what they are doing; and we can conjecture the direction in which further advances will be made. And so we can trace the development of this kind of knowledge, and in a certain and very real sense this development may be called an evolution.

But there is this difference between the evolution of nature and the evolution of the science of nature. The evolution of nature results in the existence of forms which did not exist before; the evolution of knowledge results in the perception of laws which were already in existence.

The knowledge grows, but the things known remain. The knowledge is not treated as if independent of the things known or believed to be known, as a phenomenon belonging merely to the human mind, with beginnings and laws and consequences and history of its own. And, consequently, its having a regular growth is not used as an argument against its substantial truth.

The Science of Mathematics, for instance, has a history; but no mathematician will admit that the fact that it has a history affects its claims to acceptance as truth. We may ask, how men have been brought to believe the deductions of the higher mathematics, and we may answer our own question by tracing the steps; but our conviction is not shaken that these deductions are true.

And so, too, we can trace the steps by which the great generalisations of Science have been reached, and we may show that Kepler grew out of Copernicus, and Newton out of Kepler; but the proof that the knowledge of one truth has been evolved out of the knowledge of another, and that out of the knowledge of another, is not used to show that all this Science has nothing to do with truth at all, but is only a natural growth of human thought. Science has grown through all manner of mistakes—mistakes made by the greatest thinkers and observers, mistakes which men ignorantly laugh at now, as their own mistakes will be no doubt laughed at in turn hereafter. But we do not, therefore, treat scientific thought as nothing more than one of the phenomena of humanity; ways of thinking which necessarily grew out of the conditions in which men have existed, but sufficiently accounted for by their origin and mode of growth having been shown, and having no solidity of their own.

What has been said of Science may be said also of Religion. Religion also has had its development, and in some respects a development parallel to that of Science.

It is possible to trace the steps by which men have obtained an ever larger and fuller knowledge of the Supreme Law of Right, a clearer perception of its application, of its logicalresults, of its relation to life, to conduct, to belief. It has grown through mistakes as Science has. There has been false Religion, as there has been false Science. Unsound principles of conduct have been inculcated in Religion as unsound generalisations have been set up in Science. There have been improper objects of reverence in Religion, as there have been impossible aims proposed for scientific investigation. Ezekiel rises above the doctrine that the children are punished for the sins of their parents, just as Galileo rises above the doctrine that nature abhors a vacuum. The parallel is all the more complete in that in many cases false religions have been also false sciences. The prayer to the fetish for rain is as contrary to true religion as it is contrary to true science. Many false religions are most easily overthrown by scientific instruction. Many false sciences begin to totter when the believers in them are taught true religion. The ordinary superstitions which have so strong a hold on weak characters and uninstructed minds, are as inconsistent with true faith in God as with reasonable knowledge of nature. Science grows, but the facts, whether laws or instances of the operations of those laws, are not affected by that growth. And Religion grows, but the facts of which it takes cognisance are not affected by that growth. Neither in the one case nor in the other is the fact that there has been a development any argument to show that the belief thus developed has no real foundation. The pure subjectivity of Religion, to use technical language, is no more proved by this argument than the pure subjectivity of Science.

But there is one most important particular in which the development of Religion entirely differs from the development of Science. The leaders of scientific thought, from the time that Science has been conscious of itself, have never claimed direct divine instruction. For a long time, indeed, scientific thought rested largely on tradition, and that tradition was handed on from generation to generation without any examination into its foundations. The stores of past observations seemed so very much larger in quantity than any that men could add in their own day, that it was natural to give more weight to what was received than to what was newly observed. The experience of each generation in succession seemed nothing in comparison with the accumulated experience of all preceding generations. And in many cases old traditions stopped the growth of Science by preventing the acceptance of observations inconsistent with them. But such old traditions never claimed to rest on a revelation from God; or, if such a claim was made here and there, it never had strength enough to root itself in Science and form part of the recognised authority on which Science stood.

Science, from the time when it recognised itself as Science, has owed its development to observation of nature, and long before it shook off the fetters of unexamined tradition it had disclaimed, even for that tradition, any other basis than this. But not so Religion. Many religions, and among them the purer and higher religions, in proportion to their nearer approach to perfection, have claimed to rest on a Divine Revelation, and to be something more than either speculations of philosophic observers of nature, or deductions from innate principles of reason or conscience. Not thinkers, but prophets, or men claiming to be prophets, have given the purest religions to their disciples among mankind. It has always been possible to bring all religious teaching to the bar of conscience; it has been possible to put all religious teaching to logical examination; to systematise its precepts, whether of faith or conduct; to inquire into its fundamental principles, and to ask for the authority on which the whole teaching rests. But these applications of our intellectual faculties to Religion have always been admitted as coming after, not as preceding, the teaching to which they are made. The prophet does sometimes reason when he is deducing from principles already accepted, new precepts, or new prohibitions; but he does not confine himself to such reasoning in the fulfilment of his mission. He professes to have a message to give. He will accredit it by such means as He supplies Who has sent him with this message. He will, in order to open the consciences of his hearers, appeal to past revelations which they have already received, and with which his new message is in thorough harmony; but he often appeals also to his power over nature to bear witness that the Lord of nature has sent him. The Hebrew prophet will appeal to the teaching of the Law, will repeat the old revelation with its old unshaken and unshakeable precepts, but he will not stop there: he will also give signs from the Lord to prove that he has a right to the title of prophet which he claims. Armed with this title, he will go on to predict the coming of the Great Restorer, the Messiah; he will insist on the judgment of all things, sure to be passed in its appointed day; he will hint at the immortality of the soul, and the execution of the Almighty justice on every man that lives.

It is probable enough that many of the inferior religions have grown up with no such claim at all. The worship of ancestors, where it has prevailed, has very likely, as has been suggested, grown out of dreams, in which loving memory has brought back in sleep vivid images of the dead who were reverenced while they lived, and cannot be readily forgotten after death. Such worship barely attains to what may be called in strictness a religion. Its connexion with the spiritual faculty, the true seat of religion, is weak and vague. It is like the honour paid to a sovereign residing in a distant capital, with only the difference that those who receive this worship are supposed to reside not in a distant capital, but in another world. So, too, the worship of fetishes, of trees, of serpents, of the heavenly bodies, while they have some of the inferior elements of religion in them, yet hardly deserve to be called religions. There is in them the sentiment of fear, the acknowledgment of persons or some resemblance of persons imperceptible by the senses; the acknowledgment of powers possessed by these persons. But the central idea of a rule of holiness is either altogether wanting, or so very feeble and indistinct as to contain no promise of developing into ultimate supremacy. These religions do not often lay claim to a revelation from a supreme authority. And they have withered away with the growth of knowledge and with clearer perceptions of what Religion must be if it is to exist at all.

All the higher religions have claimed to rest on a divine revelation, and the Christian Religion on a series of such revelations. The Christian Religion does not profess (as does for instance the Mahommedan) to be wrapped up in one divine communication made to one man and admitting thereafter of no modifications. Though resting on divine revelation it is professedly a development, and is thus in harmony with the Creator's operations in nature. Whether we consider what is taught concerning the heavenly Moral Law, or concerning human nature and its moral and spiritual needs, or concerning Almighty God and His dealings with us His creatures, it is undeniable that the teaching of the Bible is quite different at the end from what it is at the beginning.

The New Testament considered by itself as a body of teaching is such an advance on all that preceded it as to be quite unique in the history of the world. The ideas conveyed in the Old Testament are absorbed, transformed, completed, so as to make them as a whole entirely new; and to these are added entirely new ideas sufficient by themselves to form a whole system of doctrine. And because of this it is difficult to speak of the new teaching as having grown out of the old.

But the Old Testament covers many centuries, and within its range we can trace a steady growth, and that growth always of the same character, and always pointing towards what the Gospel finally revealed. The strength of the moral sentiment in the earlier books is always assigned to the belief in, and reverence for, Almighty God. It is evidently held to be more important to believe in God and to fear Him than to see the perfection of His holiness. If we distinguish between Religion and Morality, Religion is made the more important of the two. It is more important to recognise that the holy God exists and reigns than to see clearly in what His holiness, and indeed all holiness, consists. The sentiment of reverence is more important than the perception of that universality which we now know to be the essential characteristic of the Moral Law. In analysing the origin and nature of Religion in the second of these Lectures, it was necessary to follow the order of thought, and beginning with Duty to end with God. But the order of fact is not the same. In actual fact man began with God and ends with a clearer perception of Duty. Hence in all the earlier stages the morality is imperfect. The profaneness of Esau is a serious offence. The ungenerous temper, the unfairness and duplicity of Jacob are light in comparison. Truth is not an essential. Blood-shedding and impurity when in horrible excess are treated as most grievous sins; but restrained within limits are easily condoned. Women are placed below their true and natural place; polygamy if not distinctly allowed is certainly condoned; divorce is permitted on one side, not on the other. Slavery is allowed though put under regulation. But the unity and spirituality of God are guarded with the strongest sanctions, and nothing could be said against idolatry and polytheism now, in sterner and clearer language than was used then. The reverence for God required then was as great as the reverence required now. But the conception of the holiness which is the main object of that reverence has changed; has in fact been purified and cleared. And the change is traceable in the Old Testament. The prophets teach a higher morality than is found in the earlier books. Cruelty is condemned as it had not been before. The heathen are not regarded as outside God's love, and the future embraces them in His mercy even if the present does not. Conscience begins to be recognised and appealed to. Idolatry is not merely forbidden, its folly is exposed; it is treated not only with condemnation, but with scorn. Individual responsibility is insisted on. Children are not held responsible for their fathers, though the inheritance of moral evil and of the consequences of moral evil is never denied. And even trust in God rises to a higher level in Habakkuk's declaration that that trust shall never be shaken by any calamity that may befall him, than in the earlier belief that calamities would never befall those who held fast that trust.

If we review this progress in moral teaching we recognise that it corresponds to the natural and for the most part unconscious working of that instinctive test which, as was pointed out before, we apply to all moral questions, the test of universality. The pivots of all the prophetical teaching are the incessant inculcation of justice and mercy; justice which requiresus to recognise the rights of others side by side with our own; mercy which demands our sympathy with the feelings of other creatures that can feel.

We are bound to recognise the claims of others to equal treatment with ourselves, and any refusal or apparent refusal to do so must be justified by a universal rule applicable to all alike. The perpetual attempt to justify exceptions in this way is sure to end in diminishing the number of those exceptions. If we are compelled to think much of the position of woman in marriage, we are sure at last to come to Malachi's declaration that God hateth putting away. If we are compelled to think of the position of slaves, we cannot continue for ever to believe that there are some beings with consciences and free wills, who nevertheless, because of the accidents of their lives, have no rights at all; and we acknowledge the righteousness of Jeremiah's denunciation of the breach of covenant when the nobles of Judah re-enslaved those whom they had solemnly emancipated. If we think of the nature of responsibility and the justification of punishment, we find it impossible to believe that an innocent man shall be rightly punished for the wrong-doing of another, even if that other be his father or his mother; and we are convinced that Ezekiel is speaking God's words when he proclaims on God's behalf that 'the soul that sinneth it shall die; the son shall not bear the iniquity of the father, neither shall the father bear the iniquity of the son.' And once more, whatever divine purpose gave the chosen people a priority among all peoples in knowledge of divine will and possession of divine favour, it is impossible to find any rule by which this priority shall for ever exclude all other peoples from being within the range of God's manifested love; and conscience cannot but accept as a divine message that the Gentiles also shall come to the Heavenly 'Light, and their kings to the brightness of His rising.' So again, to turn from justice to mercy, we recognise that we are bound to spare pain to all creatures that can feel, and this duty can only be set aside by some higher duty which makes that pain the means to a higher moral end. And if we are set by our consciences to seek for some rule of universal application for this purpose, it becomes perpetually clearer that nothing can excuse cruel punishments inflicted on criminals or enemies, or hard-hearted indifference to the poor and the weak. Our own nature cries out for kindness in our pain, and that very cry from within compels our consciences to listen to the cry from without. And the denunciations of cruelty and oppression we recognise as we hear them to be the voice of God.

But however true it be that this progress corresponds exactly throughout with the necessary working of the great moral principles implanted in the spiritual faculty, it nevertheless remains true also that all this teaching in its successive stages is given by men who did not profess to be working out a philosophical system, but who claimed to bring a message from God, to speak by His authority, and in many cases to be trusted with special powers in proof of possessing that authority. Looking back over it afterwards we can see that the teaching in its successive stages was a development, but it always took the form of a revelation. And its life was due to that fact. As far as it is possible to judge, that union between Morality and Religion, between duty and faith, without which both religion and morality soon wither out of human consciences, can only be secured—has only been secured—by presenting spiritual truth in this form of a Revelation.

When we pass to the New Testament, all that has previously been taught in the Old, in so far as it is related to the new teaching at all, is related as the bud to the flower. The development, if it be indeed a development, is so great, so sudden, so strange, that it seems difficult to recognise that it is a development at all.

First, the morality is in form, if not in substance, absolutely new. The duty of justice and mercy is pushed at once to its extreme limits, even to the length of entire self-surrender. The disciple has his own rights no doubt, as every other man has his; but he is required to leave his rights in God's hands and to think of the rights of others only. The highest place is assigned to meekness in conduct and humility in spirit. The humility of the Sermon on the Mount may possibly by careful analysis be shown to be identical at bottom with the magnanimity of Aristotle's Ethics. But the presentation of the two is so utterly opposed that in the effect on life the identity is altogether lost. And as justice and mercy, so too self-discipline is pushed as far as it can go. Instead of the enjoyment of life being an integral part of the aim set before the will, hunger and thirst for righteousness, and penitence for failure in keeping to it, are to fill up the believer's hopes for himself. Of inward satisfaction and peace he is often assured; but these, and these only, are the means to that peace. The disciple's life is to consist in bearing the cross, and bearing it cheerfully; in returning good for evil, and love for indifference and even for hatred; in detaching his affections from all the pleasures to be obtained from external things; in fixing his trust and his love on his Eternal Father. Taken as a whole, this is quite unlike all moral teaching that preceded it, and there is no indication that any philosophy could ever have evolved it. It has fastened on the human conscience from the day that it was uttered; and whatever moral teaching since has not been inspired from this source has soon passed out of power and been forgotten. We find when we examine that it exactly agrees with the fundamental teaching of the spiritual faculty when that teaching is applied to such creatures as we are, and to such a God as the New Testament sets before us. But we find it impossible to assert that by any working of human thought this morality could have been obtained by the spiritual faculty unaided. On the contrary, it seems more near the truth to say that we could never have obtained so clear a conception of the great Moral Law, if the teaching of the New Testament had not enlightened and purified the spiritual faculty itself. And to this is to be added that the moral teaching of the New Testament recognises what we may now almost consider a proved necessity of our nature, or at least a sure characteristic of the government of the world, that perpetual progress without which nothing human seems to keep sweet and wholesome. Perfect as the New Testament morality is in spirit, it is nevertheless imperfect in actualprecepts. It leaves questions to be solved some of which have not been solved yet. It left slavery untouched, though assuredly doomed. It said nothing of patriotism. It gave no clear command concerning the right use of wealth. It laid down no principles for the government of states, though such principles must have a moral basis. There has been a perpetual growth in the understanding and in the application of this perfect teaching, and there will yet be a growth. Of no philosophical system of morals is it possible to say the same.

But in the second place, the New Testament contains not only a new morality, it contains also a new account of human nature. The mystery of that discord which makes the noblest and best of human souls a scene of perpetual internal conflict is acknowledged and its counterpart in God's dealings with mankind is set forth. The struggle between the spiritual faculty asserting its due supremacy, and the lower passions and appetites, impulses and inclinations, is so described by Saint Paul that none have ever since questioned his description with any effect. And our Lord's teaching of our absolute dependence on God and helplessness without Him; and Saint John's teaching that the whole world, outside Christ, 'lieth in the wicked one,' lay down the same truth. And as the mystery of moral evil in mankind is thus set forth, so too the mystery of the remedy for that evil. In the love of God shown in the Cross of Christ, in our union with God through that same Death upon the Cross is the power which conquers evil in the soul and carries a man ever upward to spiritual heights. And as all profounder thinkers have confessed the truth of the account thus given of the internal contradiction of man's moral nature, so have all believers borne witness (and only they could bear witness) to the account thus given of the solution of that contradiction and the renovation of that nature. Millions have lived and died in the Christian faith since the teaching recorded in the New Testament was given, and among them have been the purest, the justest, the most self-sacrificing, the most heavenly-minded of mankind. And they all concur in saying that the one stay of all their spiritual lives has been communion with God through Christ.

Thirdly, the New Testament affirms with a clearness previously unknown the immortality of the soul and the future gift of that spiritual body which shall in some way spring from the natural body as the plant grows from the seed. There had grown up, no doubt quite naturally, anticipations of this doctrine and ever stronger and more deeply-rooted persuasion that it must be true. But it is revealed in the New Testament as it is taught nowhere else, and it is sealed by the Resurrection of our Lord, ever since then the historical centre of the Christian Faith. How exactly it harmonises with the teaching of the spiritual faculty I have pointed out before.

And, lastly, the New Testament not only tells us what never was told before of man's nature as a spiritual being and of his destiny hereafter; it tells also what was never told elsewhere of the nature of God and of the relations between Him and His creature man. The unity and spirituality of the Godhead so strenuously insisted on in the Old Testament, is no less insisted on in the New. But the mysterious complexity embraced within that unity, though darkly hinted at in the older teaching, is nowhere clearly set forth, but in the latter. We may find anticipations of the teaching of St. Paul and St. John, and of our Lord Himself as recorded by St. John, in the Book of Proverbs, in the Prophets, in the Rabbinical writers between the Prophets and the New Testament, and we can see in Philo to what this finally came unaided by Revelation. But the Christian teaching on our Lord's nature and on the Incarnation is distinct from all this. And it is in the Christian form, and only in that form, that the doctrine has satisfied the spiritual needs of the great mass of believers.

Now there cannot be any doubt that the hold which this teaching has had upon mankind has depended entirely on the extraordinary degree in which the teaching of the Bible has satisfied the conscience. Without that no miracles however overwhelmingly attested, no external evidence of whatever kind, could have compelled intellects of the highest rank, side by side with the most uncultivated and the most barren, to accept it as divine, nor could anything else have so often rekindled its old fire at times when faith in it had apparently withered away. The teaching of the Bible has always found and must always find its main evidence within the human soul.

And the fact that the teaching of the Bible, though when examined afterwards it turns out to be development or evolution, yet was always given at the time as a revelation, so far from diminishing the force of this internal evidence adds to it still more force than it would otherwise have. For what underlies the very conception of revelation is the doctrine that all progress in higher spiritual knowledge is bound up with conscious communion with God. Now it is an experience common to all believers that in that communion is to be found not only all strength but all enlightenment also. The believer knows that he learns spiritual truth in proportion as he refers his life to God's judgment, prays to God for clearer vision of what is duty and what is right faith, and makes it his one great aim to do God's will. He uses all the faculties that God has given him to understand the great divine law; but he perpetually looks to God for instruction, and whatever else may be said of that instruction his experience tells him that his advance in spiritual knowledge is in proportion to his nearness in thought and feeling to God Himself. That the progress of the human race in spiritual knowledge, unlike progress in scientific knowledge, should be due not to thinkers intellectually gifted, but to Prophets and Apostles inspired by God, thus exactly corresponds with what the spiritually-minded man finds within his own soul. And so too does it correspond with what he sees in others. Often and often the unlearned and untrained by sheer goodness of life attain to wonderful perception of spiritual truth, and the holiness of the unlettered peasant reveals to his conscience the law of right conduct in circumstances which perplex the disciplined and well informed. As the human race has learnt the highest spiritual truth by direct communication from God, so too on communion with God far more than on intellectual power, depends the progress of spiritual knowledge in every human soul.

But though the hold of the Bible on the faith of believers unquestionably depends on its satisfying the conscience in every stage of its enlightenment, it is equally certain that those who gave the messages recorded in the Bible claimed something more as proof of their authority than the approval of the conscience of their hearers. They professed to prove their mission by the evidence of supernatural powers; and the teaching of the Bible cannot be dissociated from the miraculous element in it which is connected with that teaching. If, indeed, the Old Testament stood alone we might acknowledge that the miraculous element in it occupied comparatively so small a place, and was so separable from the rest, and the evidence for it was so rarely, if ever, contemporaneous, that it might be left out of count. But we cannot say this of the New Testament, nor in particular of the account that has reached us of the sayings and doings of our Lord. The miracles are embedded in, are indeed intertwined with, the narrative. Many of our Lord's most characteristic sayings are so associated with narratives of miracles that the two cannot be torn apart: 'I have not seen so great faith, no, not in Israel;' 'My Father worketh hitherto, and I work;' 'Son, thy sins be forgiven thee;' 'Beware of the leaven of the Pharisees and the Sadducees;' 'It is not meet to take the children's bread and cast it to dogs;' 'This kind goeth not out but by prayer and fasting;' 'Were there not ten cleansed, but where are the nine?' 'Sin no more, lest a worse thing come unto thee.' In fact, there can be no real doubt that our Lord believed that He could work miracles, and professed to work them, and that His disciples believed that He worked many, and included that fact in their meaning when they spoke of Him as going about doing good. And these disciples professed to work miracles themselves and believed that they did work them. It is of course true that they had no strictly scientific conception of a miracle, and would often have called by that name what was in reality extraordinary but not miraculous. And it is true too that, if we take each miracle by itself, there is but one miracle, namely our Lord's Resurrection, for which clear and unmistakeable and sufficient evidence is given. But while the exclusion of any one miracle as insufficiently attested is possible, the exclusion of the miraculous element altogether is not possible without a complete surrender of the position taken by the first Christian teachers. As they claimed to be inspired and to have enlightenment which was not shared by mankind at large, so did they claim, if not each for himself, yet certainly for our Lord, power not shared by ordinary men, power to step out of the ordinary course of natural events, and, whether by virtue of some higher law operative only in rare instances, or by direct interference of the Almighty, to prove a divine mission by exhibiting in fact what is an essential part of the supremacy of the Moral Law, the dominion of that Law over the physical world.

The teachers of other religions besides the Christian have claimed supernatural powers, and have professed to give a supernatural message. This is a strong evidence of the deep-seated need in the human soul for such a direct communication from God to man. Men seem to need it so much that without it they are unable to accept the truth, or to hold it long if they do accept it. All who thus claim supernatural authority must, of course, justify their claim. They must justify their message to the human conscience. What they teach must be an advance towards, and finally an expression of, the Supreme Moral Law. And if they profess to have miraculous power they must give reasonable evidence that such power is really theirs. But if they fail in this, still the fact remains that their very claim must answer to something in the spiritual nature of man, or it would not be so invariably made nor so largely successful.

It seems as if, whatever may be the ground of belief when once revelation has penetrated into the soul, the exercise of supernatural power was needed to procure that access in the first instance. We believe because we find our consciences satisfied, and we bring up our children in such discipline of conscience that they too shall have sufficient training to recognise and hold fast divine truth. And if we had lived at the time and could have had our eyes opened to see the spiritual power of the Christian Faith, we might have believed without any external evidence at all. But the first receivers of the message, to whom the revelation was new, and, as must have often happened and we actually know did happen, to whom it was hard to reconcile that revelation with previous teaching, how sure were they to need some other and outer evidence that it really came from God. The supernatural in the form of miracles can never be the highest kind of evidence, can never stand alone as evidence; but it seems to have been needed for the first reception. And there seem to be minds that need it still, and to all it is a help to find that reasonable ground can be shown for holding that such evidence was originally given.

Revelation, in short, takes a higher stand than belongs to all other teaching, and except for its having taken that higher stand it does not appear that the highest teaching would have been possible. To look back afterwards and say that we find a development or an evolution is easy. And at first sight it seems to follow that, being an evolution, it may well be no more than the outcome of the working of the natural forces. But look closer and you see the undeniable fact that all these developments by the working of natural forces have perished. Not Socrates, nor Plato, nor Aristotle, nor the Stoics, nor Philo have been able to lay hold of mankind, nor have their moral systems in any large degree satisfied our spiritual faculty. Revelation, and revelation alone, has taught us; and it is from the teaching of revelation that men have obtained the very knowledge which some now use to show that there was no need of revelation. That altruism which is now to displace the command of God is nothing but the teaching of the Sermon on the Mount robbed of its heavenly power, robbed of the great doctrine which underlies the whole sermon. For that doctrine is the Fatherhood of God which has been shown most especially in this, that from the beginning He has never forgotten His children.

LECTURE VI.

APPARENT COLLISION BETWEEN RELIGION AND THE DOCTRINE OF EVOLUTION.

'Know ye that the Lord He is God: it is He that hath made us, and not we ourselves.' *Psalm* c. 3.

Religion is rooted in our spiritual nature and its fundamental truths are as independent of experience for their hold on our consciences as the truths of mathematics for their hold on our reason.

But as a matter of fact Religion has taken the form of a revelation. And this introduces a new contact between Religion and Science, and of necessity a new possibility of collision. There is not only possible opposition or apparent opposition of Science in what is revealed, in what we may call the actual substance of the revelation; but also in the accessories and evidences of the revelation, which may be no actual part of the revelation itself, but nevertheless are, to all appearance, inseparably bound up with it. It is therefore no more than might have been expected that the general postulate of the uniformity of nature should appear to be contravened by the claim to supernatural power made on behalf of revelation, and that the special, but just at present leading scientific doctrine, the doctrine of Evolution, should be found inconsistent with parts, or what appear to be parts, of the revelation itself. And we have to consider the two questions, What has Revelation to say concerning Evolution? and what has Science to say concerning Miracles?

Concerning Evolution, we have first to consider how much in this direction has been made fairly probable, and what still remains to be determined.

It cannot then be well denied that the astronomers and geologists have made it exceedingly probable that this earth on which we live has been brought to its present condition by passing through a succession of changes from an original state of great heat and fluidity, perhaps even from a mixture mainly consisting of gases; that such a body as the planet Jupiter represents one of the stages through which it has passed, that such a body as the moon represents a stage toward which it is tending; that it has shrunk as it cooled, and as it shrank has formed the elevations which we call mountains, and the depressions which contain the seas and oceans; that it has been worn by the action of heat from within and water from without, and in consequence of this action presents the appearance when examined below the surface of successive strata or layers; that different kinds of animal and vegetable life have followed one another on the surface, and that some of their remains are found in these strata now; and that all this has taken enormous periods of time. All this is exceedingly probable, because it is the way in which, as Laplace first pointed out, under well-established scientific laws of matter, particularly the law of gravitation and the law of the radiation of heat, a great fluid mass would necessarily change. And the whole solar system may and probably did come into its present condition in this way. It certainly could have been so formed, and there is no reason for supposing that it was formed in any other way.

Once more, if we begin, as it were, at the other end, and trace things backwards from the present, instead of forwards from the remote past, it cannot be denied that Darwin's investigations have made it exceedingly probable that the vast variety of plants and animals have sprung from a much smaller number of original forms.

In the first place, the unity of plan which can be found pervading any great class of animals or plants seems to point to unity of ancestry. Why, for instance, should the vertebrate animals be formed on a common plan, the parts of the framework being varied from species to species, but the framework as a whole always exhibiting the same fundamental type? If they all descended from a common ancestor, and the variations were introduced in the course of that descent, this remarkable fact is at once accounted for. But, in the second place, observation shows that slight variations ARE perpetually being introduced with every successive generation, and many of these variations are transmitted to the generations that follow. In the course of time, therefore, from any one parent stock would descend a very large variety of kinds. But if, in the third place, it be asked why this variety does not range by imperceptible degrees from extreme forms in one direction to extreme forms in the other, the answer is to be found in the enormous prodigality and the equally enormous waste of life and living creatures. Plants and animals produce far more descendants than ever come even to such maturity as to reproduce their kind. And this is particularly the case with the lower forms of life. Eggs and seeds and germs are destroyed by millions, and so in a less but still enormous proportion are the young that come from those that have not been destroyed. There is no waste like the waste of life that is to be seen in nature. Living creatures are destroyed by lack of fit nourishment, by lack of means of reproduction, by accidents, by enemies. The inevitable operation of this waste, as Darwin's investigation showed, has been to destroy all those varieties which were not well fitted to their surroundings, and to keep those that were. One species of animal has been preserved by length of neck, which enabled it to reach high-growing fruits and leaves; another by a thicker skin, which made it difficult for enemies to devour; another by a colour which made it easier to hide. One plant has been preserved by a bright flower which attracted insects to carry its pollen to other flowers of its kind; another by a sweet fruit which attracted birds to scatter its seed. Meanwhile other animals and plants that had not these advantages perished for the lack of them. The result would be to maintain, and perpetually, though with exceeding slowness, more and more to adapt to the conditions of their life, those species whose peculiarities gave them some advantage in the great struggle for existence.

Here again we have the working of known laws of life, capable of accounting for what we see. And the high probability cannot be denied that by evolution of this kind the present races of living creatures have been formed. And to these arguments the strongest corroboration is given by the frequent occurrence, both in plants and animals, of useless parts which still remain as indications of organs that once were useful and have long become useless. Animals that now live permanently in the dark have abortive eyes which cannot see, but indicate an ancestor with eyes that could see. Animals that never walk have abortive legs hidden under their skin, useless now but indicating what was useful once. Our knowledge no doubt in this as in any other province of nature is but the merest fraction of what may be known therein. But there is no evidence whatever to show that what we have observed is not a fair sample of the whole. And so taking it, we find that the mass of evidence in favour of the evolution of plants and animals is enormously great and increasing daily.

Granting then the high probability of the two theories of Evolution, that which begins with Laplace and explains the way in which the earth was fitted to be the habitation of living creatures, and that which owes its name to Darwin and gives an account of the formation of the living creatures now existing, we have to see what limitations and modifications are necessarily attached to our complete acceptance of both.

First, then, at the very meeting point of these two evolutions we have the important fact that all the evidence that we possess up to the present day negatives the opinion that life is a mere evolution from inorganic matter. We know perfectly well the constituents of all living substances. We know that the fundamental material of all plants and all animals is a compound called protoplasm, or that, in other words, organic matter in all its immense variety of forms is nothing but protoplasm variously modified. And we know the constituent elements of this protoplasm, and their proportions, and the temperatures within which protoplasm as such can exist. But we are quite powerless to make it, or to show how it is made, or to detect nature in the act of making it. All the evidence we have points to one conclusion only, that life is the result of antecedent life, and is producible on no other conditions. Repeatedly have scientific observers believed that they have come on instances of spontaneous generation, but further examination has invariably shown that they have been mistaken. We can put the necessary elements together, but we cannot supply the necessary bond by which they are to be made to live. Nay, we cannot even recall that bond when it has once been dissolved. We can take living protoplasm and we can kill it. It will be protoplasm still, so far as our best chemistry can discover, but it will be dead protoplasm, and we cannot make it live again; and as far as we know nature can no more make it live than we can. It can be used as food for living creatures, animals or plants, and so its substance can be taken up by living protoplasm and made to share in the life which thus consumes it; but life of its own it cannot obtain. Now here, as it seems, the acceptance of the two evolutions lands us in acceptance of a miracle. The creation of life is unaccounted for. And it much more exactly answers to what we mean by a miracle than it did under the old theory of creation before Evolution was made a scientific doctrine. For under that old theory the creation of living creatures stood on the same footing as the creation of metals or other inorganic substances. It was part of that beginning which had to be taken for granted, and which for that reason lay outside of the domain of Science altogether. But if we accept the two evolutions, the creation of life, if unaccounted for, presents itself as a direct interference in the actual history of the world. There could have been no life when the earth was nothing but a mass of intensely heated fluid. There came a time when the earth became ready for life to exist upon it. And the life came, and no laws of inorganic matter can account for its coming. As it stands this is a great miracle. And from this conclusion the only escape that has been suggested is to suppose that life came in on a meteoric stone from some already formed habitable world; a supposition which transfers the miracle to another scene, but leaves it as great a miracle as before.

Nor, if it was a miracle, can we deny that there was a purpose in it worthy of miraculous interference. For what purpose can rank side by side with the existence and development of life, the primary condition of all moral and spiritual existence and action in this world? In the introduction of life was wrapped up all that we value and all that we venerate in the whole creation. The infinite superiority, not in degree only, but in kind, of the living to the lifeless, of a man to a stone, justifies us in believing that the main purpose of the creation that we see was to supply a dwelling-place and a scene of action for living beings. We cannot say that the dignity of the Moral Law requires that creatures to be made partakers in the knowledge of it, and even creatures of a lower nature but akin to them, must have been the results of a separate and miraculous act of creation. But we can say that there is a congruity in such a miracle, with the moral purpose of all the world, of which we are a part, that removes all difficulty in believing it. Science, as such, cannot admit a miracle, and can only say, 'Here is a puzzle yet unsolved.' Nor can the most religious scientific man be blamed as undutiful to religion if he persists in endeavouring to solve the puzzle. But he has no right to insist beforehand that the puzzle is certainly soluble; for that he cannot know, and the evidence is against him.

Secondly, if we look at the Darwinian theory by itself, we see at once that it is incomplete, and the consideration of this incompleteness gravely modifies the conclusion which would otherwise be rightly drawn from it, and which, indeed, Darwin himself seems disposed to draw. For the theory rests on two main pillars, the transmission of characteristics from progenitor to progeny, and the introduction of minute variations in the progeny with each successive generation. Now, the former of these may be said to be well established, and we recognise it as a law of life that all plants and animals propagate their own kind. But the latter has, as yet, been hardly examined at all. Each new generation shows special slight variations. But what causes these variations? and what determines what they shall be? In Darwin's investigations these questions are not touched. The variations are treated as if they were quite indefinite in number and in nature. He concerns himself only with the effect of these variations after they have appeared. Some have the effect of giving the plant or animal an advantage in the struggle of life; some give no such advantage; some are hurtful. And hence follows the permanent preservation or speedy destruction of the plants and animals themselves. But we are bound to look not only to their effects but to their causes, if the theory is to be completed. And then we cannot fail to see that these variations in the progeny cannot be due to something in the progenitors, or otherwise the variations would be all alike, which they certainly are not. They must, therefore, be due to external circumstances. These slight variations are produced by the action of the surroundings, by the food, by the temperature, by the various accidents of life in the progenitors. Now, when we see this, we see also how gravely it modifies the conclusions which we have to draw concerning the ancestry of any species now existing. Let us take, for instance, the great order of vertebrate animals. At first sight the Darwinian theory seems to indicate that all these animals are descended from one pair or one individual, and that their unity of construction is due to that fact; but if we go back in thought to the time at which the special peculiarities were introduced which really constituted the order and separated it from other animals, we see that it is by no means clear that it originated with one pair or with one individual, and that, on the contrary, the probabilities are the other way. Although the separation of this order from the rest must have taken place very early, it cannot well have taken place until millions of animals had already come into existence. The prodigality of nature in multiplying animal life is fully acknowledged by Darwin, and that prodigality is apparently greatest in the lowest and most formless type of animal. There being, then, these many millions of living creatures in existence, the external surroundings introduce into them many variations, and among these the special variations to which the vertebrate type is due. It is quite clear that wherever the external surroundings were the same or nearly the same, the variations introduced would be the same or nearly the same. Now, it is far more probable that external surroundings should be the same or nearly the same in many places than that each spot should be absolutely unlike every other spot in these particulars. The beginnings of the vertebrate order would show themselves simultaneously, or at any rate independently, in many places wherever external conditions were sufficiently similar. And the unity of the plan in the vertebrata would be due, not to absolute unity of ancestry, but to unity of external conditions at a particular epoch in the descent of life. Hence it follows that the separation of animals into orders and genera and even into species took place, if not for the most part yet very largely, at a very early period in the history of organic evolution. Of course the descendants of any one of the original vertebrata might, and probably in not a few cases did, branch off into new subdivisions and yet again into further subdivisions, and we are always justified in looking for unity of ancestry among all the species. But it is also quite possible that any species may be regularly descended, without branching off at all, from one of the originals, and that other species that resemble it may owe the resemblance simply to very great similarity of external conditions. To find, for instance, the unity of ancestry between man and the other animals, it will certainly be necessary to go back to a point in the history of life when living creatures were as yet formless, undeveloped—the materials, as we may call them, of the animal creation as we now see it, and not in any but a strictly scientific sense, what we mean when we ordinarily speak of animals. The true settlement of such questions as these can only be obtained when long and patient study shall have completed

Darwin's investigations by determining under what laws and within what limits the slight variations which characterise each individual animal or plant are congenitally introduced into its structure. As things stand the probabilities certainly are that a creature with such especial characteristics as man has had a history altogether of his own, if not from the beginning of all life upon the globe, yet from a very early period in the development of that life. He resembles certain other animals very closely in the structure of his body; but the part which external conditions had to play in the earliest stages of evolution of life must have been so exceedingly large that identity or close similarity in these external conditions may well account for these resemblances. And the enormous gap which separates his nature from that of all other creatures known, indicates an exceedingly early difference of origin.

Lastly, it is quite impossible to evolve the Moral Law out of anything but itself. Attempts have been made, and many more will no doubt be made, to trace the origin of the spiritual faculty to a development of the other faculties. And it is to be expected that great success will ultimately attend the endeavours to show the growth of all the subordinate powers of the soul. That our emotions, that our impulses, that our affections should have had a history, and that their present working should be the result of that history, has nothing in it improbable. There can be no question that we inherit these things very largely, and that they are also very largely due to special peculiarities of constitution in each individual. That large part of us which is rightly assigned to our nature as distinct from our own will and our own free action, it is perfectly reasonable to find subject to laws of Evolution. Much of this nature, indeed, we share with the lower animals. They, too, can love; can be angry or pleased; can put affection above appetite; can show generosity and nobility of spirit; can be patient, persevering, tender, self-sacrificing; can take delight in society: and some can even organise it, and thus enter on a kind of civilisation. The dog and the horse, man'sfaithful servants and companions, show emotions and affections rising as far as mere emotions and affections can rise to the human level. Ants show an advance in the arts of life well comparable to our own. If the bare animal nature is thus capable of such high attainments by the mere working of natural forces, it is to be expected that similar forces in mankind should be found to work under similar laws. We are not spiritual beings only, we are animals, and whatever nature has done for other animals we may expect it to have done and to be doing for us. And if their nature is capable of evolution, so too should ours be. And the study of such evolution of our own nature is likely to be of the greatest value. This nature is the main instrument, put into the grasp as it were of that spiritual faculty which is our inmost essence, to be used in making our whole life an offering to God. It is good to know what can be done with this instrument and what cannot; how it has been formed in the past, and may be still further formed for the future. It is good to study the evolution of humanity. But all this does not touch the spiritual faculty itself, nor the Moral Law which that faculty proclaims to us. The essence of that law is its universality; and out of all this development, when carried to its very perfection, the conception of such universality cannot be obtained. Nothing in this evolution ever rises to the height of a law which shall bind even God Himself and enable Abraham to say, 'Shall not the Judge of all the earth do right?' The very word right in this, its fulness of meaning, cannot be used.

Evolution may lead the creature to say what is hateful and what is loveable, what is painful and what is delightful, what is to be feared and what is to be sought; it may develope the sentiment which comes nearest of all to the sentiment of reverence, namely, the sentiment of shame; but it cannot reveal the eternal character of the distinction between right and wrong. Nay, there may be, as was pointed out in the last Lecture, an evolution in our knowledge even of the Moral Law, just as there is an evolution in our knowledge of mathematics. The fulness of its meaning can become clearer and ever clearer as generation learns from generation. But the principle of the Moral Law, its universality, its supremacy, cannot come out of any development of human nature any more than the necessity of mathematical truth can so come. It stands not on experience, and is its own evidence. Nor indeed have any of the attempts to show that everything in man (religion included) is the product of Evolution ever touched the question how this conception of universal supremacy comes in. It is treated as if it were an unauthorised extension from our own experience to what lies beyond all experience. This, however, is to deny the essence of the Moral Law altogether: that Law is universal or it is nothing.

Now, when we compare the account of the creation and of man given by the doctrine of Evolution with that given in the Bible, we see at once that the two are in different regions. The purpose of giving the accounts is different; the spirit and character of the accounts is different; the details are altogether different. The comparison must take note of the difference of spirit and aim before it can proceed at all.

It is then quite certain, and even those who contend for the literal interpretation of this part of the Bible will generally admit, that the purpose of the revelation is not to teach Science at all. It is to teach great spiritual and moral lessons, and it takes the facts of nature as they appear to ordinary people. When the creation of man is mentioned there is clearly no intention to say by what processes this creation was effected, or how much time it took to work out those processes. The narrative is not touched by the question, Was this a single act done in a moment, or a process lasting through millions of years? The writer of the Book of Genesis sees the earth peopled, as we may say, by many varieties of plants and animals. He asserts that God made them all, and made them resemble each other and differ from each other. He knows nothing and says nothing of the means used to produce their resemblances or their differences. He takes them as he sees them, and speaks of their creation as God's work. Had he been commissioned to teach his people the science of the matter, he would have had to put a most serious obstacle in the way of their faith. They would have found it almost impossible to believe in a process of creation so utterly unlike all their own experience. And it would have been quite useless to them besides, since their science was not in such a condition as to enable them to coordinate this doctrine with any other. As science it would have been dead; and as spiritual truth it would have been a hindrance.

But he had, nevertheless, great ideas to communicate, and we can read them still.

He had to teach that the world as we see it, and all therein contained, was created out of nothing; and that the spiritual, and not the material, was the source of all existence. He had to teach that the creation was not merely orderly, but progressive; going from the formless to the formed; from the orderless to the ordered; from the inanimate to the animate; from the plant to the animal; from the lower animal to the higher; from the beast to the man; ending with the rest of the Sabbath, the type of the highest, the spiritual, life. Nothing, certainly, could more exactly match the doctrine of Evolution than this. It is, in fact, the same thing said from a different point of view. All this is done by casting the account into the form of a week of work with the Sabbath at the end. In so constructing his account, the writer made use of a mode of teaching used commonly enough in the Bible. The symbolical use of the number seven is common in other inspired writers. The symbolical use of periods of time is not without example. That the purpose of the account was not to teach great truths, but to give men information upon scientific questions, is incredible. And, in fact, if we look in this account for literal history, it becomes very difficult to give any meaning to what is said of the seventh day, or to reconcile the interpretation of it with our Lord's words concerning the Sabbath, 'My Father worketh hitherto, and I work.' There is no more reason for setting aside Geology, because it does not agree in detail with Genesis, than there is for setting aside Astronomy because all through the Old Testament the sun is spoken of as going round the earth.

And when the writer of Genesis passes from creation in general to man in particular, it is still clear that he has no mission to tell those for whom he was writing by what processes man was formed, or how long those processes lasted. This was as alien from his purpose as it would have been to tell what every physiologist now knows of the processes by which every individual man is developed from a small germ to a breathing and living infant. He takes men—and he could not but take men as he sees them—with their sinful nature, with their moral and spiritual capacity, with their relations of sex, with their relations of family. He has to teach the essential supremacy of man among creatures, the subordination in position but equality in nature of woman to man, the original declension of man's will from the divine path, the dim and distant but sure hope of man's restoration. These are not, and cannot be, lessons of science. They are worked out into the allegory of the Garden of Eden. But in this allegory there is nothing whatever that crosses the path of science, nor is it for reasons of science that so many great Christian thinkers from the earliest age of the Church downwards have pronounced it an allegory. The spiritual truth contained in it is certainly the purpose for which it is told; and evolution such as science has rendered probable had done its work in forming man such as he is before the narrative begins.

It may be said that it seems inconsistent with the dignity of man's nature as described in the Bible to believe that his formation was effected by any process of evolution, still more by any such process of evolution as would represent him to have been an animal before he became a man.

But, in the first place, it is to be observed that Science does not yet assert, and there is no reason to believe that it ever will assert, that man became a fully developed animal, with the brute instincts and inclinations, appetites and passions, fully formed, an animal such as we see other animals now, before he passed on into a man such as man is now. His body may have been developed according to the theory of Evolution, yet along a parallel but independent line of its own; but at any rate it branched off from other animals at a very early point in the descent of animal life. And, further, as Science cannot yet assert that life was not introduced into the world when made habitable by a direct creative act, so too Science cannot yet assert, and it is tolerably certain will never assert, that the higher and added life, the spiritual faculty, which is man's characteristic prerogative, was not given to man by a direct creative act as soon as the body which was to be the seat and the instrument of that spiritual faculty had been sufficiently developed to receive it. That the body should have been first prepared, and that when it was prepared the soul should either have been then given, or then first made to live in the image of God,—this is a supposition which is inconsistent neither with what the Bible tells nor with what Science has up to this time proved.

And to this must be added that it is out of place for us to define what is consistent or inconsistent with the dignity of man in the process or method by which he was created to be what he is. His dignity consists in his possession of the spiritual faculty, and not in the method by which he became possessed of it. We cannot tell, we never can tell, and the Bible never professes to tell, what powers or gifts are wrapped up in matter itself, or in that living matter of which we are made. How absolutely nothing we know of the mode by which any single soul is created! The germ which is to become a man can be traced by the physiologist through all the changes that it has to undergo before it comes to life. Is the future soul wrapped up in it from the first, and dormant till the hour of awakening comes? or is it given at some moment in the development? We see in the infant how its powers expand, and we know that the spiritual faculty, the very essence of its being, has a development like the other faculties. It has in it the gift of speech, and yet it cannot speak. Judgment, and taste, and power of thought; self-sacrifice and unswerving truth; science and art, and spiritual understanding, all may be there in abundant measure and yet may show no sign. All this we know; and because it is common and well known we see nothing inconsistent with the dignity of our nature in this concealment of all that dignity, helpless and powerless, within the form of an infant in arms. With this before us it is impossible to say that anything which Science has yet proved, or ever has any chance of proving, is inconsistent with the place given to man in Creation by the teaching of the Bible.

In conclusion, we cannot find that Science, in teaching Evolution, has yet asserted anything that is inconsistent with Revelation, unless we assume that Revelation was intended not to teach spiritual truth only, but physical truth also. Here, as in all similar cases, we find that the writer of the Book of Genesis, like all the other writers in the Bible, took nature as he saw it, and expressed his teaching in language corresponding to what he saw. And the doctrine of Evolution, in so far as it has been shown to be true, does but fill out in detail the declaration that we are 'fearfully and wonderfully made; marvellous are Thy works; and that my soul knoweth right well.' There is nothing in all that Science has yet taught, or is on the way to teach, which conflicts with the doctrine that we are made in the Divine Image, rulers of the creation around us by a Divine superiority, the recipients of a Revelation from a Father in Heaven, and responsible to judgment by His Law. We know not how the first human soul was made, just as we know not how any human soul has been made since; but we know that we are, in a sense in which no other creatures living with us are, the children of His special care.

LECTURE VII.

APPARENT COLLISION OF SCIENCE WITH THE CLAIM TO SUPERNATURAL POWER.

'Believe Me that I am in the Father, and the Father in Me: or else believe Me for the very works' sake.' *St. John* xiv. 11.

Science and Religion come into apparent collision on the question of the freedom of the will. Science and Revelation come into a similar apparent collision on the possibility of miracles. The cases are precisely parallel. In each individual man the uniformity of nature is broken to leave room for the moral force of the will to assert its independent existence. This breach of uniformity is within very narrow limits, and occurs much more rarely than appears at first sight. But the demand to admit not only the possibility but the fact of this breach is imperative, and to deny it is to turn the command of the Moral Law as revealed in the conscience into a delusion. So, too, Revelation asserts its right to set aside the uniformity of nature to leave room for a direct communication from God to man. It is an essential part of the Divine Moral Law to claim supremacy over the physical world. Unless somehow or other the moral ultimately rules the physical, the Moral Law cannot rightly claim our absolute obedience. Revelation as given to us maintains that this superiority has been asserted in fact here in the world of phenomena. To deny this is very nearly equivalent to denying that any revelation has been made. In this way Revelation asserts, for God's message to the human race precisely the same breach of uniformity which every man's conscience claims for himself in regard to his own conduct.

It is, however, necessary to point out that when we speak of a breach of uniformity we are never in a position to deny that the breach of uniformity may be physical only and perhaps apparent only. It may be found, it probably will be found, at last that the Moral Law has in some way always maintained its own uniformity unbroken. The Moral Law has in its essence an elasticity which the physical law has not. It often takes the form, that, given certain conduct, there will follow certain consequences; and the law is kept though the conduct is free. It is further possible, and Revelation has no interest in denying it, that the intervention which has apparently disturbed the sequence of phenomena is, after all, that of a higher physical law as yet unknown. For instance, the miraculous healing of the sick may be no miracle in the strictest sense at all. It may be but an instance of the power of mind over body, a power which is undeniably not yet brought within the range of Science, and which nevertheless may be really within its domain. In other ways what seems to be miraculous may be simply unusual. And it must therefore be always remembered that Revelation is not bound by the scientific definition of a miracle, and that if all the miraculous events recorded in the Bible happened exactly as they are told, and if Science were some day able to show that they could be accounted for by natural causes working at the time in each case, this would not in any way affect their character, as regards the Revelation which they were worked to prove or of which they form a part. Revelation uses these events for its own purposes. Some of these events are spoken of as evidences of a divine mission. Some of them are substantive facts embraced in the message delivered. And if for these purposes they have served their turn, if they have arrested attention which would not otherwise have been arrested, if they have overcome prejudices, if they have compelled belief, the fact that they are afterwards discovered to be no breach of the law of uniformity has no bearing at all on the Revelation to which they belong. The miracle would in that case consist in the precise coincidence in time with the purpose which they served, or in the manner and degree in which they marked out the Man who wrought them from all other men, or in the foreshadowing of events which are in the distant future.

Thus, for instance, it is quite possible that our Lord's Resurrection may be found hereafter to be no miracle at all in the scientific sense. It foreshadows and begins the general Resurrection; when that general Resurrection comes we may find that it is, after all, the natural issue of physical laws always at work.

There is nothing at present to indicate anything of the sort; but a general resurrection in itself implies not a special interference but a general rule. If, when we rise again, we find that this resurrection is and always was a part of the Divine purpose, and brought about at last by machinery precisely the same in kind as that which has been used in making and governing the world, we may also find that our Lord's Resurrection was brought about by the operation of precisely the same machinery. We may find that even in the language of strict science 'He was the first fruits of them that slept,' and that His Resurrection was not a miracle, but the first instance of the working of a law till the last day quite unknown, but on that last day operative on all that ever lived.

Let us compare the general resurrection with the first introduction of life into the world. As far as scientific observation has yet gone that first introduction of life was a miracle. No one has ever yet succeeded in tracing it to the operation of any known laws. If it is a miracle it is a miracle precisely similar in kind to the miracle which believers are expecting at the last day. And assuredly if a miracle was once worked to introduce life into this habitable world, there is very good reason to expect that another miracle will be worked hereafter to restore life to those that have lived. But there are scientific men who think that the introduction of life was not a miracle, that it came at the fitting moment by the working of natural laws; or, in other words, that such properties are inherent in the elements of which protoplasm is made that in certain special circumstances these elements will not only combine but that the product of their combination will live. If this be so, it is assuredly no such very strange supposition that there may be such properties inherent in our bodies or in certain particles, whether particles of matter or not, belonging to our bodies, that in certain special circumstances these particles will return to life. And if this be so the general resurrection may be no miracle, but the result of the properties originally inherent in our bodies and of the working of the laws of those properties. And as the general resurrection so our Lord's

Resurrection may in this way turn out to be no breach of the uniformity of nature.

But this new discovery, if then made, would not affect the place which our Lord's Resurrection holds in the records of Revelation. It is not the purpose of Revelation to interfere with the course of nature; if such interference be needless, and the work of revealing God to man can be done without it, there is no reason whatever to believe that any such interference would take place.

Or, take again any of our Lord's miracles of healing. There is no question at all that the power of mind over body is exceedingly great, and has never yet been thoroughly examined. We know almost nothing of the extent of this power, of its laws, of its limits. Marvellous recoveries often astonish the physician, and he cannot account for them except by supposing that in some way the powers of the mind have been roused to interfere with the working of the nervous system. And some men, on the other hand, have died or their health has been shattered by mere imaginations. Some men of note have attributed the recoveries claimed for homœopathy to this cause. Some have assigned to this cause the extraordinary cures that have been undeniably wrought at the shrines, or on sight or touch of the relics, of Roman Catholic saints. The different impostures that have on many occasions prevailed for a time and then lost their reputation and passed out of fashion, are generally supposed to have owed their short-lived success to the same obscure working of unknown natural laws. They have been tested by their successes and their failures. They have succeeded, and for a time continued to succeed; but at last they have ceased to work because faith in them for some reason or other has been shaken down. Their falsehood has thus been detected; but nevertheless their genuine success for a time has been enough to show that they rested on a reality, and that reality seems to have consisted in the strange power of mind over body. In this region all is at present unexamined; and all operations are tentative, and for that reason most are only successful for a time. Now our Lord's miracles are never tentative; that is not the character given to them either by friend or by foe. Nor is there any instance recorded either by friend or by foe of an attempted miracle not accomplished. Nowhere is there any record given us by the assailants of the Gospel of any instance of His action parallel to the record given in the Acts of the Apostles of the seven sons of Sceva the Jew. The accounts of his enemies charge Him with deceit, which is identical with saying that they did not believe Him. But they do not ever charge Him with failure. Nevertheless it is quite conceivable that many of His miracles of healing may have been the result of this power of mind over body which we are now considering. It is possible that they may be due not to an interference with the uniformity of nature, but to a superiority in His mental power to the similar power possessed by other men. Men seem to possess this power both over their own bodies and over the bodies of others in different degrees. Some can influence other men's bodies through their minds more; some less. Possibly He may have possessed this power absolutely where others possessed it conditionally. He may have possessed it without limit; others within limits. If this were so, these acts of healing would not be miracles in the strictly scientific sense. They would imply very great superiority in Him to other men. But they would be in themselves under the law of uniformity. Now it is clear that if this should turn out to be so, though these acts would not be miracles for the purposes of Science, they would still be miracles for the purposes of Revelation. They would do their work in arresting attention, and still more in accrediting both the message and the Messenger. They would separate Him from ordinary men. They would prove Him to be possessed of credentials worth examining. To the believer it would make no difference whether Science called them miracles or not. To him it would still remain the fact that here was a Messenger whom God had seen fit to endow with powers which no other man ever possessed in such degree and such completeness, though others may have possessed some touch of them greater or less.

Further, it is necessary to repeat what was briefly remarked in a previous Lecture, that the position which miracles take as regards us who read them many centuries after, and as regards those who witnessed and recorded them at the time, is quite different. To them the miracles were the first and often the chief proof that the man who wrought them had been sent by God, and that His message was a revelation, not an imposture; to us they are, if accepted at all, accepted as a part of the revelation itself. There are no doubt a few minds that are convinced by Paley's argument, and beginning with accepting the miracles as proved by sufficient external evidence, go on to accept the conclusion that therefore the teaching that was thus accompanied must be divine. But most men are quite unable to take to pieces in this way the records in which Revelation is contained, and to go from external evidence taken alone to the messengers who thus proved their mission, and thence to the substance of the message which they taught. To most of us, on the contrary, the Revelation is a whole, capable of being looked at from many sides, and found to be divine from whatever side it is seen; and one of its aspects is this supernatural character by which it appears to assert its identity with that Moral Law which claims absolute supremacy over all the physical world. The main evidence of the Revelation to us consists in its harmony with the voice of the spiritual faculty within us; and the claim which it asserts to have come through teachers endowed with supernatural power is so far corroborative evidence as it falls in with the essential character of the Moral Law. That eternal law claims supremacy over the physical world and actually asserts it in the freedom of the human will; and a Revelation which comes from Him Who in His own essential Being is that very law personified, might be expected to exhibit the same claim in actual manifestation in its approach to men.

Bearing these limitations and characteristics of the miraculous element in the Bible in mind, let us ask how that miraculous element is therein presented.

First, in the account of the creation, it is taught that the original existence of all matter flows from a spiritual source. We do not define God as the cause, meaning that that is His essence, and that except as causing other things to exist He does not exist Himself. But we describe Him as the Cause, meaning that all things exist by His Will, and that without His Will nothing could ever have existed. And as the Revelation tells us that He is the source of all existence, the Creator of the substance of things, so too does it assert that He gave all things their special properties and the laws of those properties, and that not only the original creation, but all the subsequent history of all things has been the outcome of His design, and that He has thus prescribed the government of the whole universe. And yet again the Revelation from the beginning to the end maintains His living Presence in and over all things that He has thus formed, and denies that He has parted with His power to do fresh acts of creation, fresh acts of government, whenever and wherever He sees fit. For He is necessarily free and cannot be restrained by anything but His own holiness. And unless He expressly revealed to us that His own holiness prevented Him from interfering with His own creation, we could not put limits to what He could do. The Revelation that He has given us says just the contrary, and from end to end implies that He is present in the government of the creation which He has made.

What evidence, then, is there in the world of phenomena that He has ever thus interfered? Putting aside as untenable all idea of *a priori* impossibility, admitting that God can work a miracle if He will, admitting that a miracle avowedly worked in the interest of a divine revelation stands on a totally different footing from a miracle avowedly worked in any other interest, putting the breach of the law of uniformity made by a miracle on the same footing as the breach of the same law made by a human will; we have to ask what evidence can be given that any such miracles as are recorded in the Bible have ever been worked?

It is plain at once that the answer must be given by the New Testament. No *such* evidence can now be produced on behalf of the miracles in the Old Testament. The times are remote; the date and authorship of the Books not established with certainty; the mixture of poetry with history no longer capable of any sure separation into its parts; and, if the New Testament did not exist, it would be impossible to show such a distinct preponderance of probability as would justify us in calling on any to accept the miraculous parts of the narrative as historically true.

But in the New Testament we stand on different ground. And we have here first the evidence which Paley has put together to show that the early Christians spent their lives and finally surrendered their lives as witnesses to a Gospel which included miracles both among its evidences and as part of its substance. It is not possible to get rid of miracles nor the belief in miracles from the history of the Apostles. They testify to our Lord's Resurrection as to an actual fact, and make it the basis of all their preaching. They testify to our Lord's miracles as part of the character of His life. It is necessary to maintain that they were mere fanatics with no claim to respect but rather to the pity which we feel for utterly ignorant goodness, if we are to hold that no miracle was ever wrought by our Lord. It is difficult to maintain even their honesty if they preached the Resurrection of our Lord without any basis of fact to rest on. No man who is not determined to uphold an opinion at all hazards can question that St. Paul and St. Peter believed that our Lord rose from the dead, and that they died for and in that belief.

But, in the second place, behind the Apostles stands our Lord Himself, and whatever may be said of the documents that compose the New Testament, they are at any rate sufficient to show that our Lord was universally believed by His disciples to have the power of working miracles and to have often worked them. There is no hesitation in regard to this; no hint of any doubt. But not only so, there is no hint of any disclaimer on His part. He must have known whether He could work miracles or not. He must have known that His disciples believed Him to possess the power. There is not the slightest trace of His ever having implied that this was a misconception. He did sometimes disclaim what was ascribed to Him, even when what was ascribed to Him was truly His, but was ascribed to Him without real knowledge of what it implied. 'Why callest thou Me good? There is none good but One, that is, God,' we DO find. But 'Why askest thou Me to do this? There is none that can do this but One, that is God,' we do NOT find. It is plain that He accepted the belief that assigned Him powers above those of other men—powers given Him by His Father in heaven—and never discouraged it. Nay, He demanded it. Take the lowest ground, and admit for argument's sake that the New Testament contains a legendary element, and still you cannot cut the miracles out of the Gospels and Epistles without altering them beyond recognition. The Jesus Christ presented to us in the New Testament would become a different person if the miracles were removed. And if He claimed to possess and exercise this power, the evidence becomes the evidence of One Who must have known and Whom we cannot disbelieve.

And this claim, which He has thus made, and which was thus accepted by His disciples, is corroborated by the power, different in form but similar in kind, which He exerted then on the men of His own day, and has ever since continued to exert on all succeeding generations. The first disciples were under His absolute dominion. They preached Christ and not themselves. They referred everything to Him, and professed to have no power but from Him. St. Paul with all his genius and marvellous power of influence, yet professes to be nothing without Christ and to be everything in Christ. Our Lord left no writing behind Him, but committed His Revelation to His Apostles, and we only know Him through them. But they are not like ordinary disciples of a great teacher; philosophers succeeding a philosopher; prophets succeeding a prophet. To no one of them does it occur for a moment to teach anything except as from Him. St. Paul gives advice sometimes which he does not profess to be giving by our Lord's command, but when he does so, he puts the mark of his own inferiority on what he says, and claims for it no such authority as belongs to a word from Christ. A word from Christ was final on all subjects.

And this power over men has never weakened from that day to this. There is no other power like it in the world. Science proceeds in far the majority of cases by trial of some theory as a working hypothesis. Such too has been the procedure of Christian Faith. Trust Christ; stake your happiness on Him; stake your hope of satisfying all spiritual aspirations on Him; stake your power of winning the victory over temptation on Him—this is the exhortation of Apostle, and martyr, and saint, and evangelist, and pastor, and teacher. And those who have thus tried the strength of the Christian hypothesis have not failed. The Christian Church has been stained with many a blot. Ill deeds have been wrought in the name of Christ. Evil laws have been passed. Strange superstitions have prevailed. But no other body can show such saints, no other body can produce so great a cloud of witnesses. It is certain that the lives and the deaths, the characters and the aims, of those who have trusted their all to Christ have made them what He bade them be, the salt of the earth. And they testify with one voice that they know no other power which has upheld them but the power of Christ whom they have taken for their Lord. Others have sometimes been set up as in some sort rivals to Him as teachers or as examples; but here there is no rival even pretended. In no other man have men been called on to believe as a living present power, able to give strength and victory in the conflicts of the soul. The Church, too, has passed through times of spiritual depression, we may almost say of degradation. And in the worst of times within the Church there has always remained a wonderful recuperative power, which has shaken off inconsistencies and defects in the past, and will do so yet more in the future. But this recuperative power has always shown itself in one form, and in one form only, namely, a return to Christ and to trust in Him, a trust which has never been falsified.

The martyrdom of our Lord's disciples is enough to prove that belief in His supernatural powers and in His exercise of those powers was no gradual growth of later times, but from the very beginning rooted in the convictions of those who must have known the truth. The character of our Lord as revealed in the Gospels makes it impossible to disbelieve His claims whatever they may be. His power attested by generations of believers ever since corroborates those claims by the persistent evidence of eighteen centuries.

Against this evidence what is to be said?

It is said that the evidence for the uniformity of nature is so overwhelming that nothing can set it aside. And further it is said, that, even if it be conceded that it might be set aside, no evidence sufficient for the purpose has yet been produced.

Now to deal with this second assertion first, we must ask what is the nature of the evidence that would be deemed sufficient? If the inquirer does not believe that God created and still governs the world, assuredly no evidence will ever be sufficient to convince him that God has worked a miracle. The existence of God is certainly not to be proved by His interference with nature. Had He desired to reveal Himself to us primarily in that way, He would have wrought many more miracles than we now know of, and would have kept our faith alive by perpetual and unmistakeable manifestations of His presence and power. But He has not so willed. He has made our belief in Him rest mainly on the voice within ourselves, in order that we might walk by faith and not by sight. It will be a hopeless task to convince men that there is a God by pointing, not to His creation but to His interference with creation. But if a man do believe there is a God, what kind of evidence ought he to expect to show him that God has interfered in the course of the creation?

In the first place, he must not expect that the physical evidence, that is the miraculous evidence, for Revelation should be of such a character as to stand above the spiritual evidence. Just as the fundamental evidence for the existence of a God is to be found in the voice of conscience, and the arguments from design and from the order and beauty and visible purpose of the creation are secondary—corroborative not demonstrative—so too the primary evidence of a Revelation from God must be found in the harmony of that Revelation with the voice of conscience, and only the secondary and corroborative evidence is to be looked for in miracles. And in both cases the reason is the same. For it is not God's purpose to win the intellectually gifted, the wise, the cultivated, the clever, but to win the spiritually gifted, the humble, the tender-hearted, the souls that are discontented with their own shortcomings, the souls that have a capacity for finding happiness in self-sacrifice. It would defeat the purpose of the Revelation made to us if the hard-headed should have an advantage in accepting it over the humble-minded. The evidence must be such that spiritual character shall be an element in the acceptance of it. There would be a contradiction, if the faculty whereby we mainly recognised God were the spiritual faculty, and the faculty whereby we mainly recognised His Revelation were the scientific faculty.

And, in the second place, we have no right to expect that the evidence for miracles wrought in one age should be such evidence as properly belongs to another age. It is sometimes urged that the evidence supplied by the testimony of the early Christians is of little value because it was never cross-examined. No such precautions surrounded the evidence as would now be required to give any value to evidence of similar events. The witnesses gave up their lives to attest what they taught; but there was no one to scrutinise what they asserted. St. Paul's evidence on our Lord's Resurrection cannot now be put to the test of searching questions. But to make such objections as these is to make what is on the face of it an absurd demand. It is to ask that the scientific processes of the nineteenth century should have been anticipated in the first, that men should be miraculously guided to supply a kind of evidence which would be utterly superfluous at the time in order to be convincing eighteen hundred years afterwards. This would indeed have put the miraculous incidents in the New Testament narrative altogether out of place, and made the miracles more important than the Revelation which they were worked to introduce.

Now, if these two conditions are borne in mind, it is difficult to see what better evidence could be obtained of a miraculous life than we possess concerning the life of our Lord.

The moral and spiritual evidence is His own character which intentionally overshadows all the rest, and it is inconceivable that He should have made a false claim. And the material evidence is the testimony of men who freely gave their lives in proof of what they said. Nor has anything yet been said or written to shake Paley's argument on this point.

But, if we pass on to the other objection, that no evidence can ever be sufficient to prove a miracle because the evidence for the uniformity of nature is so overwhelming, we can only see in such an assertion an instance of that inability to get out of an accustomed groove against which Science has perpetually to guard. In Science the uniformity of nature is so indispensable a postulate, that without it we cannot stir a step. And if the student of Science is to admit a breach, it can only be by stepping outside of his science for the time and conceiving the possibility that there is some other truth beside scientific truth, and some other kind of evidence beside scientific evidence. We have all heard of the need of guarding against the bondage in which custom binds the mind. We have heard of the student who when first he saw a locomotive looked perseveringly for the horses that impelled it, because he had never known, and consequently could not imagine any other mode of producing such motion. But this danger attends not only the separate investigations which Science makes into phenomena; it attends Science as a whole. And it is necessary repeatedly to insist on the fact that Science has not proved and cannot prove that the scientific domain is co-extensive with nature itself.

The evidence for the uniformity of nature consists in the fact that from the beginning of Science the known reign of physical law has been steadily extending without a check; that instance after instance of apparent exception has been brought by further examination within its province; that the hypothesis of uniformity has now been long on trial and has never yet been found to fail; that no one who has so tried it has the slightest hesitation in trusting it for the future, as he has proved it in the past. But clearly as this evidence proves a general, it never gets beyond a general, uniformity. It has not succeeded in showing that the human will comes under the same rule. It has not succeeded in silencing the voice within us, which claims superiority for the moral over the physical. And when the utmost extent of human knowledge is compared with the vastness of nature, the claim to extend the induction from generality to universality is seen to be utterly untenable. So much as this, indeed, Science has rendered highly probable, that the uniformity of nature is never broken except for a moral purpose. It is only for such a purpose that the will is ever free. It is only for such a purpose that Revelation has ever claimed to be superior to nature. But beyond this Science cannot go. Let it be granted that the claim for freedom of the will has been often unduly pushed far beyond this limit, and let it be granted that religions professing to be revelations have included records of miracles which had no moral purpose. This does not affect the general conclusion that the evidence for uniformity has never succeeded, and can never succeed in showing, that the God who made and rules the universe never sets aside a physical law for a moral purpose, either by working through the human will or by direct action on external nature.

Science will continue its progress, and as the thoughts of men become clearer it will be perpetually more plainly seen that nothing in Revelation really interferes with that progress. It will be seen that devout believers can observe, can cross-question nature, can look for uniformity and find it, with as keen an eye, with as active an imagination, with as sure a reasoning, as those who deny entirely all possibility of miracles and reject all Revelation on that account. The belief that God can work miracles and has worked them, has never yet obstructed the path of a single student of Science; nor has any student who repudiated that belief found any aid in his study from that repudiation. The rush of Science of late years has for the time made many men fancy that Science is everything; and believers in Revelation have helped this fancy by insisting on their part that Revelation is everything; but such waves of opinion, resting really on feeling, are sure to pass away, and scientific men will learn that there are other kinds of knowledge besides scientific knowledge, as believers are already learning that God teaches us by other methods besides the method of Revelation. The students of the Bible will certainly learn that Revelation need not fear the discoveries of Science, not even such doctrines as that of Evolution. And the students of nature will certainly learn that Science has nothing to fear from the teaching of Revelation, not even from the claim to miraculous power. For most certainly both Science and Revelation come from one and the same God; 'the heavens declare His glory, and the firmament showeth His handywork; His law is perfect, converting the soul; His testimony is sure, making wise the simple.'

LECTURE VIII.

THE CONCLUSION OF THE ARGUMENT.

'No man can say that Jesus is the Lord, but by the Holy Ghost.' 1 *Cor.* xii. 3.

It is now the proper time to review the argument of these Lectures, and to endeavour to trace, if possible, the source of the estrangement which just at present separates Religion and Science.

The postulate of Science is admitted on all hands to be the uniformity of nature, and the proof of this postulate has been found to consist in an induction from the facts which nature presents and our senses observe. Uniformity is quickly noticed, and after it has been noticed for some time it is instinctively used as a working hypothesis. So used it accumulates perpetually increasing evidence of its truth, and if we except two great classes of facts, we never find any instance of its failure. The two classes of facts which are thus excepted are the acts of the human will and the miraculous element in Revelation, both of them instances of one thing, namely, the interference of the moral with the physical. To complete the induction and to deprive the denial of universal uniformity of all evidence to rest on, all that is necessary is to get rid of these two exceptions. If Science could get rid of these exceptions, though it could not be said that the fundamental postulate was demonstrated, it could be said that all the evidence was in its favour and absolutely no evidence against it. And although scientific belief would then still rank below mathematical belief, it would nevertheless have a cogency quite irresistible. Science would not thereby gain in power of progress, in practical acceptance, or in utility to man. But men are so constituted that completeness gives a special kind of satisfaction not to be got in any other way. If Science could but be complete it would seem to gain in dignity, if it gained in nothing else. And it is easy to foster a kind of passion for this completeness until every attempt to question it is resented. I have seen a boy first learning mechanics show a dislike to consider the effect of friction as marring the symmetry and beauty of mechanical problems; too vague, too uncertain, too irregular to be allowed any entrance into a system which is so rounded and so precise without it. And something of the same temper can sometimes be seen in students of Science at the very thought of there being anything in the world not under the dominion of the great scientific postulate. The world which thus contains something which Science cannot deal with is pronounced forthwith to be not the world that we know, not the world with which we are concerned; a conceivable world if we choose to indulge our imagination in such dreams, but not a real world either now or at any time before or after. And yet the freedom of the human will and the sense which cannot be eradicated of the responsibility attaching to all human conduct, perpetually retorts that this world in which we live contains an element which cannot be subdued to obedience to the scientific law, but will have a course of its own. The sense of responsibility is a rock which no demand for completeness in Science can crush. All attempts at reconciling the mechanical firmness of an unbroken law of uniformity with the voice within that cannot be silenced telling us that we must answer for our action, have failed, and we know that they will for ever fail.

If indeed it could be said that the progress of Science was really barred by this inability to make the induction complete, and to assert the unbroken uniformity of all nature; if it could be said that any uncertainty was thus cast over scientific conclusions, or any false or misleading lights thus held up to draw inquirers from the true path, it would undoubtedly become a duty to examine, and to examine anxiously, whether indeed it could be true that our faculties were thus hopelessly at variance with each other, the scientific faculty, imposing on us one belief, and the spiritual faculty another, and the two practically irreconcileable. But there is no reason whatever for thinking this. Newton's investigations were unquestionably pursued, as all true scientific investigations must ever be pursued, in reliance on the truth of the uniformity of nature, and yet he never felt it the slightest hindrance to his progress that he always tacitly and often expressly acknowledged that God had reserved to Himself the power of setting this uniformity aside, and indeed believed that He had used this power. The believer who asserts the universality of a law except when God works a miracle to set it aside is certainly at no real disadvantage in comparison with an unbeliever who makes the same assertion with no qualification at all. It is granted on all hands that miracles are, and ever have been, exceedingly rare, and for that reason need not be taken into account in the investigation of nature. It is granted that the freedom of the human will works within narrow limits, and very slowly and slightly affects the great mass of human conduct and what depends on human conduct. And Science has often to deal with approximations when nothing but approximations can be obtained. We perpetually meet in nature with quantities and relations that cannot be accurately expressed nor accurately ascertained, and we have to be content with approximations, and we know how to use them in Science. Many chemical properties can only be so expressed; many primary facts, such as the distances, the volumes, the weights of heavenly bodies; and yet the approximations serve our purpose. And so too, if there be a reserve still uncovered by the scientific postulate, that will not in any degree affect our investigation of what is so covered.

In short, the unity of all things which Science is for ever seeking will be found not in the physical world alone, but in the physical and spiritual united. That unity embraces both. And the uniformity which is the expression of that unity is not a uniformity complete in nature, taken by itself, but complete when the two worlds are taken together. And this Science ought to recognise.

Let us turn from the physical to the spiritual.

The voice within us which demands our acceptance of religion makes no direct appeal to the evidence supplied by the senses. We are called on to believe in a supreme law of duty on pain of being lowered before our own consciences. And this law of duty goes on to assert its own supremacy over all things that exist, and that not as an accidental fact, but as inherent in its essence. And this supremacy cannot be other than an accidental fact unless it be not only actual but intended. And intention implies personality; and the law thus shows itself to be a Supreme Being, claiming our reverence, and asserting Himself to be the Creator, the Ruler, and the Judge of all things that are. And this same voice within us asserts that we are responsible to Him for all our conduct, and are capable of that responsibility because free to choose what that conduct shall be. We are to believe not because the truth of this voice is proved independently of itself, but simply because we are commanded. Corroborative evidence may be looked for elsewhere, but the main, the primary evidence is within the soul.

Hence the strength of this belief depends on ourselves and on our own character. To every man the voice speaks. But its authority is felt in proportion to the spirituality of each who hears. Its acceptance is bound up in some way with our own wills. How far it is a matter of choice to believe or to disbelieve it is not possible to define. The will lies hidden as it were behind the emotions, the affections, the nobler impulses. The conscience shades off into the other faculties, and we cannot always isolate it from the rest. But though it be impossible to say precisely how the will is concerned in the spiritual belief, there can be no doubt that it always takes its part in such belief. It is the keen conscience, it is the will that can be moved to its depths by the conscience, that grasp most strongly the certainty of the law of duty. It is the man with the strongest and noblest aspirations, the man who sees the beauty of humility, the man who feels most strongly the deep peace of self-sacrifice, *that* is the man who finds the voice within most irresistible. It is not by any means always the man who lives the most correct life; correctness of life may be due to natural and not to spiritual causes. And the man whom we should find faultless in point of morals may yet be wanting in spiritual depth, and not have as yet, and perhaps may not have to the last, the spiritual faculty strong within him. But the man, even if he have many and grievous faults, who nevertheless is keenly susceptible of higher things, is the one to whom the voice within speaks with authority not to be gainsaid, and to him that voice is final.

It is this fact that the perception of things spiritual varies from man to man, and depends on character, and involves action of the will, that makes it always possible to represent our knowledge of the law of duty as in itself standing on a less sure foundation than our knowledge of scientific truth. Whether a man has or has not the necessary power of mind to comprehend scientific reasoning is tested with comparative ease. And if he have that power, the reasoning is certain in course of time to be understood, and when it is understood it compels assent so long as it keeps within its own proper domain. But the perception of spiritual truth depends on a faculty whose power or weakness it is far more difficult to test; and it involves the will which may be exerted on either side. And for this reason men sometimes dismiss this truth as being no more than an imagination, needed by some men to satisfy an emotional nature, but having no substance that can be brought to an external test. The believer in God knows that the truth which he holds is as certain as the axioms of mathematics; but he cannot make others know this whose spiritual faculty is not awake; and he is liable to be asked for proof not of the spiritual but of the physical kind.

Now this much must be acknowledged, that we cannot but expect the claim to supremacy over all things to show itself in some way in the creation which has come from Him who makes that claim. It would, no doubt, be a serious difficulty if things physical and things spiritual were cut off from one another by an absolute gulf; if we were required to believe that God had created and now ruled everything, and yet we could trace not the slightest evidence of His hand either in the creation or in the history of the world.

There are then two ways in which we are able to recognise Him even in this world of phenomena. For in the first place, the creation in its order and its beauty and its marvellous adaptation of means to ends, confirms the assertion of the spiritual faculty that it owes its origin to an intelligent and benevolent purpose, exhibited in the form in which purpose is always exhibited. It works towards ends which we should expect a holy and benevolent Creator to have in view, and it accomplishes those ends in so large a proportion that, making allowance for the limited range of our knowledge, the general aim of the whole is seen with sufficient clearness. The argument is not strong enough to compel assent from those who have no ears for the inward spiritual voice, but it is abundantly sufficient to answer those who argue that there cannot be a Creator because they cannot trace His action. And the scientific doctrine of Evolution, which at first seemed to take away the force of this argument, is found on examination to confirm it and expand it. The doctrine of Evolution shows that with whatever design the world was formed, that design was entertained at the very beginning and impressed on every particle of created matter, and that the appearances of failure are not only to be accounted for by the limitation of our knowledge, but also by the fact that we are contemplating the work before it has been completed.

And in the second place, while the creation, the more closely it is examined the more distinctly shows the marks of the wisdom and goodness of the Creator, so the history of the world exhibits in the Revelation made to man clear proofs of that heavenly love which corresponds to the character of Him who has put love at the head of all the requirements of His law. The Revelation given to us has undeniably made a real mark on the world. It has upheld millions of men in a holiness of life corresponding in a very real degree to the holiness required by the law of duty. It has perpetually more and more cleared up the true teaching of that law. It is still continuing the same process, and generation after generation is better able to understand that teaching. Its fruits have been a harvest of saints and martyrs, some known and reverenced, some quite unnoticed. It has leavened all literature and all legislation. It has changed the customs of mankind and is still changing them. And if it be replied that all this is nothing but one form of the development of humanity and shows no proof of a Divine Ruler, we have a right to ask what then could be the source of such a development, and how is it that so great a power should always have worked in the name of God and should have always referred everything to His command? That fanaticism should plead God's authority without any right to do so is intelligible. But is it intelligible that all this truth and justice and purity and self-sacrificing love, all this obedience to the Supreme Law, should be the fruit of believing a lie? If there be a God, it is to be expected that He would communicate with His creatures if those creatures were capable of receiving the communication; and if He did communicate with His creatures it is to be expected that His communication would be such as we find in the Bible. The purpose of the Bible, the form of it, the gradual formation of it, the steadily-growing Revelation contained in it, these harmonise with the moral law revealed originally in the conscience. And the effect which the Revelation has produced on human history is real and great. The power which God's Revelation has exerted on the world is an undeniable fact among phenomena. It is not a demonstration of His existence; but it is a full answer to those who say, 'If God made and rules the world why do we find no signs of His hand in its course?'

And thirdly, this Revelation has not merely taken the form of a message or a series of messages, but has culminated in the appearance of a person who has always satisfied and still satisfies the conception formed by our spiritual faculty of a human representation of the divine law. Our Lord's life is that law translated into human action, and all the more because human faculties had not first framed the conception which He then came to fulfil, but He exhibited the ideal, and our conception rose as it were to correspond to it. And, as He includes in Himself all the teaching, so does He give from Himself all the power of the Revelation which He came to crown. And every true disciple of Christ can bear witness to the reality of that power in sustaining the soul.

Thus has the God, whom our spiritual faculty commands us to worship and to reverence, shown Himself in the world of phenomena. And He has given proofs of His existence andHis character precisely corresponding to the conception which He has enabled, and indeed commanded, us to form of Him. And it is because the proofs that He has given are of this nature that we are tempted to ask for more proofs of a different kind.

For it is undeniable that believers and unbelievers alike are perpetually asking for proofs that shall have more of the scientific and less of the religious character, proofs that shall more distinctly appeal to the senses. Believers in all ages have longed for external support to their faith; unbelievers have refused to believe unless supplied with more physical evidence. Believers shrink from being thrown inwards on themselves; they fear the wavering of their own faith; they are alarmed at the prospect of the buttresses of their belief being taken from them. They find it easier to believe the spiritual evidence, if they can first find much physical evidence. They wish (to use the Apostle's words) to walk by sight and not by faith. And unbelievers want a tangible proof that shall compel their understanding before it awakes their conscience. They demand a Revelation, not only confirmed by miracles at the time, but confirmed again and again by repeated miracles to every succeeding generation. They want miracles in every age adapted to the science of the age, miracles which no hardness of heart would be able to deny, which would convince the scientific man through his Science independently of his having any will to make holiness his aim when he had been convinced. This kind of evidence it has not pleased God to give. It is not the scientific man that God seeks as such, any more than it is the ignorant man that He seeks as such.

And the proofs that He gives are plainly in all cases conditioned by the rule that the spiritually minded shall most easily and most keenly perceive their force.

And, as far as unbelievers are concerned, I do not see that more need be said except to tell them that this rule is inflexible, and that it is by another way that they must look to find God, and not by the way that they insist on choosing. But believers who are in the same case need to be warned of some very real dangers that always attend a faith which makes too much of things not spiritual.

For, first, there is a real and great danger that the spiritual may be altogether obscured by the literal and the physical. We look back with astonishment on the Rabbinical interpretations of the Old Testament, and all the more because of the really great and true thoughts that are sometimes to be found in the midst of their fanciful conceits. We can trace the mischief they did to true Religion by the perverted reverence with which they regarded the words and even the letters, and the very shapes of the letters, in which their sacred books were written. Their perversions of the law of God, their subtle refinements of interpretation, their trivial conceits, their false and misleading comments and inferences, all certainly tended to encourage the hypocrisy which our Lord rebuked, and against which St. Paul contended. But we still see something of the same spirit in the attempt to maintain a verbal and even literal inspiration of the whole Bible, filling it not with the breath of a Divine Spirit, but with minute details of doctrine and precept often questionable, and, whenever separated from the principles of the eternal law, valueless or even mischievous. God's Word, instead of leading us to Him, is made to stand between and hide His face.

But, secondly, there is a serious risk that if the mind be fastened on things external in some way connected with, but yet distinct from the substance of Revelation, it may turn out that these external things cannot hold the ground on which they have been placed. They have to be given up by force at last, when they ought to have been given up long before. And when given up they too often tear away with them part of the strength of that faith of which they had previously been not only the buttress outside but a part of the living framework. It is distinctly the fault of religious, not of scientific men, that there was once a great contest between the Bible and Astronomy, that there has since been a great contest between the Bible and Geology, that there is still a great contest between the Bible and Evolution. In no one of these cases was the Revelation contained in the Bible in danger, but only the interpretation commonly put on the Bible. It is easy long afterwards to condemn the opponents of Galileo and speak of their treatment of him and his teaching as fanaticism and bigotry; and such condemnation has not unfrequently been heard from the very lips that nevertheless denounced the teaching of the geologists. But in all these cases the principle has been the same, and believers have insisted that the Bible itself was gone unless their interpretation of it was upheld. And the mischief is double. For many believers, and more especially unlearned believers, instead of gently helping one another to form the necessary modification of their view of the Bible teaching, instead of endeavouring to find the way out of the perplexity and to disentangle the true spiritual lesson from the accessories which are no part of itself, insisted that it must be all or nothing, and prepared for themselves a very severe trial. There was no doctrine involved whatever; there was nothing at stake on which the spiritual life depended. The duty to be patient, to enquire carefully, to study the other side, to wait for light, was as plain as any duty could be. But all this was forgotten in a somewhat unreasoning impulse to resist an assault on the faith. And there cannot be a doubt that on all these occasions many believers have been seriously shaken by slowly finding out that the position they have taken is untenable. When men have to give up in such circumstances they generally give up far more than they need, and in some cases an unreasonable resistance has been followed by an equally unreasonable surrender. And while believers have thus prepared a stumblingblock for themselves they have put quite as great a stumblingblock before others. For students of Science, informed by instant voices all around that they must choose between their Science and the Bible, knowing as they did that their Science was true, and supposing that the lovers and defenders of the Bible best knew what its teaching was, had no choice as honest men but to hold the truth as far as they possessed it and to give up the Bible in order to maintain their Science. It was a grievous injury inflicted on them; and though some among them might deserve no sympathy, there were some whom it was a great loss to lose.

But in the third place, the result of this clinging to externals is to shut out Science and all its correlative branches of knowledge from their proper office of making perpetually clearer the true and full meaning of the Revelation itself. It is intended that Religion should use the aid of Science in clearing her own conceptions. It is intended that as men advance in knowledge of God's works and in power of handling that knowledge, they should find themselves better able to interpret the message which they have received from their Father in Heaven. Our knowledge of the true meaning of the Bible has gained, and it was intended that it should gain, by the increase of other knowledge. Science makes clearer than anything else could have made it the higher level on which the Bible puts what is spiritual over what is material. I do not hesitate to ascribe to Science a clearer knowledge of the true interpretation of the first chapter of Genesis, and to scientific history a truer knowledge of the great historical prophets. The advance of secular studies, as they are called, clears up much in the Psalms, and much in the other poetical Books of Scripture. I cannot doubt that this was intended from the beginning, and that as Science has already done genuine service to Religion in this way, so will it do still better service with process of time.

On this side also, as on the scientific side, the teaching of the spiritual faculty and the teaching of Revelation indicate that the physical and the spiritual worlds are one whole, and that neither is complete without the other. Science enters into Religion, and is its counterpart, and has its share to take in the conduct of life and in the formation of opinion. And the believer is bound to recognise its value and make use of its services.

In conclusion, it is plain that the antagonism between Science and Religion arises much more from a difference of spirit and temper in the students of each than from any inherent opposition between the two. The man of Science is inclined to shut out from consideration a whole body of evidence, the moral and spiritual; the believer is inclined to shut out the physical. And each, from long looking at that evidence alone which properly belongs to his own subject, is inclined to hold the other cheap, and to charge on those who adduce it either blindness of understanding or wilful refusal to accept the truth. And when such a conflict arises it is the higher and not the lower, it is Faith and not Science that is likely to suffer. For the physical evidence is tangible, and the perception of it not much affected by the character of the man who studies it; the spiritual evidence stands unshaken in itself, but it is hid from eyes that have no spiritual perception, and that perception necessarily varies with the man.

By what means then can a man keep his spiritual perception in full activity? And is there any test by which a man may know whether his spiritual faculty is in contact with the source of all spiritual life and is deriving from that source the full flow of spiritual power? Revelation, if it tells us anything, ought to tell us this. And the answer which Revelation makes is expressed in the words of St. Paul, 'No man can say that Jesus is the Lord, but by the Holy Ghost.' This doctrine runs through the New Testament, and it implies that one main purpose of our Lord's appearance among men was to give them in His life, His character, His example, His teaching, at once a touchstone by which they could always try their own spirits, and judge of the real condition of their own spiritual faculty, and also a vivid presentation of the supreme spiritual law by which they could for ever more and more elevate and purify and strengthen their own spiritual power and knowledge.

Let a man study the Jesus of the Gospels. Let him put before his *conscience* the teaching that Jesus gives; the picture drawn of our Father in Heaven whose holiness cannot allow a stain upon a single soul, and whose tenderness cannot endure that a single soul should perish; Who ruleth all the universe, and yet without whom not a sparrow falleth to the ground; the picture drawn of the ideal human life, the humility, the hunger and thirst after righteousness, the utter self-sacrifice, the purity; the picture drawn of human need, the helplessness, the hopelessness of man without God. Let him ponder on all this and on the many touching expressions, the truth, the depth, the force, the superhuman sweetness and gentlenesswith which all is presented. And if his conscience bows before it, and can say without reserve and in unalloyed sincerity, 'This is my Lord; He shall be my teacher; here I recognise the fulness of the eternal law; at His feet will I henceforth sit and learn; through Him will I drink of the well-springs of eternal truth; His voice will I trust to the very utmost;' then may that man be sure that his conscience is in contact with the Father of spirits, and that his study will guide him into fuller and clearer knowledge, and more certain conviction that he is grasping the truth of God.

Let a man put before his *heart* our Lord's own character. Let him think of the life of privation without complaint, of service to His kind without a thought of self; of His unfailing sympathy with the unhappy, of His tenderness to the penitent; of His royal simplicity and humility; of His unwearied perseverance in the face of angry opposition; of His deep affection for the friends of His choice even when they deserted Him in His hour of darkness; of His death on the Cross and the unearthly love that breathed in every word Heuttered and everything He did. Let him read all this many times; and if his heart goes out to the Man whom he is thus beholding, if he can say with all his soul, This is my Lord; here is the supreme object of my affection; Him will I love with all my strength; from Him I will never, if I can help it, let my heart swerve; no other do I know more worthy to be loved; no other will I keep more steadily before my eyes; no other will I more earnestly desire to imitate; no other shall be my example, my trust, my strength, my Saviour; if a man can say this, it is certain that his heart is touched by God, and the heavenly fire is kindled in his soul.

Let a man put before his *will* the Lord's commands; the aims, the self-restraints, the aspirations that the Lord required in His disciples. Let him ponder on the call to heavenly courage in spite of all that earth can inflict or can take away; the call to take up the Cross and follow Him that was crucified; the warnings and the promises, the precepts and the prohibitions; let him think of the Leader who never flinched, of the Lawgiver who outdid His own law; let him think on the nobleness of the aims to which He pointed; of the promise of inward peace made to those who sacrificed themselves, made by our Lord and re-echoed from the very depths of our spiritual being; let him think of the sure help promised in return for absolute trust, tried by millions of saints and never yet known to fail. Let a man put this before his will, and if he can say with all his soul, This is my Lord; here I recognise Him who has a right to my absolute obedience; here is the Master that I mean to serve and follow; and in spite of my own weakness and blindness, in spite of my sins, in spite of stumbling and weariness of resolution, in spite of temptations and in spite of falls, I will not let my eyes swerve, nor my purpose quit my will; through death itself I will obey my Lord and trust to Him to carry me through whatever comes; that man most certainly is moving in the strength of God, and the power of the Eternal Spirit lives within him.

Our Lord is the crown, nay, the very substance of all Revelation. If He cannot convince the soul, no other can. The believer stakes all faith on His truth; all hope on His *power*. Ifthe man of Science would learn what it is that makes believers so sure of what they hold, he must study with an open heart the Jesus of the Gospels; if the believer seeks to keep his faith steady in the presence of so many and sometimes so violent storms of disputation, he will read of, ponder on, pray to, the Lord Jesus Christ.

Milton Keynes UK
Ingram Content Group UK Ltd.
UKHW050654020524
442115UK00011B/308